The Rightful Place of Science:

Creative Nonfiction

The Rightful Place of Science:
Creative Nonfiction

Edited by
Michael L. Zirulnik, Lee Gutkind,
and David Guston

Contributors
Elizabeth Popp Berman
Adam Briggle
Ross Carper
Roberta Chevrette
David Guston
Lee Gutkind
Gwen Ottinger
Angela Records
Sonja Schmid
Meera Lee Sethi
Sara Whelchel
Rachel Zurer

Consortium for Science, Policy & Outcomes
Tempe, AZ and Washington, DC

THE RIGHTFUL PLACE OF SCIENCE:
Creative Nonfiction

For information on the Rightful Place of Science series,
write to: Consortium for Science, Policy & Outcomes
PO Box 875603, Tempe, AZ 85287-5603

The Rightful Place of Science series explores the complex interactions among science, technology, politics, and the human condition.

Other volumes in this series:

Sarewitz, D., ed. 2014. *The Rightful Place of Science: Government & Energy Innovation*. Tempe, AZ: Consortium for Science, Policy & Outcomes.

Pielke, Jr., R. 2014. *The Rightful Place of Science: Disasters & Climate Change*. Tempe, AZ: Consortium for Science, Policy & Outcomes.

Alic, J. A. 2013. *The Rightful Place of Science: Biofuels*. Tempe, AZ: Consortium for Science, Policy & Outcomes.

Zachary, G. P., ed. 2013. *The Rightful Place of Science: Politics*. Tempe, AZ: Consortium for Science, Policy & Outcomes.

ISBN: 0692366156

ISBN-13: 978-0692366158

LCCN: 2015900584

FIRST EDITION, FEBRUARY 2015

CONTENTS

PREFACE

Kevin Finneran

Science is the realm of experiment, data, mathematical proof, and the ultimate triumph (illusion?) of rational objectivity. Public policy, particularly in these days of legislative gridlock, is the realm of ideas, values, and the ultimate triumph of ideological consistency. The interaction of these two world views is supposed to produce the Platonic ideal of evidence-based, principle-driven science policy.

The problem with this chimera is that these two world views do not readily mix because they are so fundamentally different. And even if some grand synthesis could be formed, it would find it difficult to take root in human culture and institutions because it still incorporates only two of the many perspectives through which humans try to make sense of their world.

The effort is not pointless. At least I hope it is not pointless because I and many others have devoted our careers to this work. We do need to search for a way to integrate these two powerful perspectives. But the endpoint is not a meeting of these two types of minds. Nor is it new legislation, a shift in federal funding, or an executive order. The goal is to create a culture and institutions that include in their DNA an understanding of the need to meld values with science and social goals with physical realities. The catch is that this type of genetic engineering cannot be achieved with the analytic tools that are used to achieve the intellectual synthesis. This form of synthetic biology can be accomplished only with stories.

This is not news. Everyone has discovered the power of narrative. The problem here is that most have embraced only the simplest notion of narrative. Scientists are enamored of the relentless struggle of brilliant men—and a few odd women—who with their indomitable will and rigorous respect for evidence have expanded the frontiers of knowledge. Ideologues love anecdotes that are really little parables that provide pictures of a world that neatly conforms with their world view. But these stories capture only a tiny slice of life, just as scientific and ideological analyses capture only two of the many intellectual frameworks for understanding reality.

True narrative aims to do much more; it aspires to portray reality in all its richness as it happens at the level of individuals. Stories are interesting because they are not about stick-figure stereotypes of the pointy-headed professor and caricatures of the blustering ideologue. They are about individuals in all their splendid inconsistencies and surprises. They acknowledge that scientists are not always rational, that partisan politicians are not indifferent to data, and that the majority of people do not think like scientists or policymakers.

The making of public policy is a story, not a scientific paper or an essay on political philosophy. In a democratic society, policy is created through a social process that must be acceptable to an incredibly heterogeneous voting public. And once established, it will succeed as intended only if it is responsive to the way that individuals and institutions actually function in this culture. The best way to understand policy in context is through stories that open our minds to the incongruous, the irrational, the idiosyncratic, the serendipitous, and the tragic that lie in wait to ambush our evidence-based, values-informed dreams. In that sense, stories are the route to public policy that can survive reality, that is resilient.

And the reality that narrative nonfiction writers must face is that this is an impossible task. The world is too messy, knowledge is too limited, time is too short. They are doomed to fail. But what an ambitious and glorious failure! Lee Gutkind and David Guston have issued the challenge and lighted the path for those willing to take the leap into writing stories meant to inform and inspire readers to tackle the problems that can be caused, exacerbated, or eased by science and technology. We should all be grateful to the young scientists and writers who have accepted the challenge and are blazing the trail to a more inclusive, more grounded, more nuanced, and more engaging way to write about science policy. This is not an easy road, but it is the most potentially fruitful route to a more inclusive, informed, principled, and practical science policy.

1

TO THINK, TO WRITE, TO PUBLISH

Lee Gutkind, David Guston, and Gwen Ottinger

To Think

In which two different kinds of people have to learn to think together under difficult circumstances and do things that while not utterly unprecedented are still rare and challenging.

Sonja Schmid knows that she has only three minutes to make her point—and she has to share that time with her partner. Ross Carper is standing behind her. He's in his early 30s, balding, in a striped jersey, and reading over her shoulder as she follows her notes. She describes a cigarette break she shared with an aging Russian nuclear reactor operator she had been interviewing which led to a special moment, when their conversation went beyond technical talking points to a personal topic—his relationship with his reactors.

Schmid, with a cascade of wavy black hair, black-rimmed glasses and fashionable red scarf, is an assistant professor in the Department of Science and Technology in Society at Virginia Tech University. Carper is a writer with the Environmental Molecular Sciences Laboratory, at the Department of Energy's Pacific Northwest National Laboratory. Schmid has a Ph.D. in Science & Technology Studies and Carper an MFA in fiction writing. They met each other only a few hours ago and yet they were already collaborating on a writing project that would consume them for 18 months. Now they had but three minutes to convince four very critical editors that readers—educated people who had rarely, if ever, given thought to nuclear reactors or their operators—would *want* to read about Schmid's research. Their secret weapon: the ancient art of storytelling, embodied in a genre called creative nonfiction.

This was the pitch slam—focal event of "To Think, To Write, To Publish," a project supported by the U.S. National Science Foundation. NSF usually funds research in the natural sciences or engineering to advance both fundamental knowledge and socially beneficial innovation. But NSF is also interested in two activities in conjunction with these goals: funding scholarship that helps explore and explain the ethical, social, and policy dimensions of science and innovation—what we call science and innovation policy, or SIP; and communicating to broad public audiences the importance of science and innovation. It is rare, however, for NSF to support projects that attempt to do both—to bring together scholars of science and innovation policy (like Schmid) and communicators of science and innovation (like Carper). The idea behind the project was that, by working with communicators, scholars could render their academic findings more cinematic, introduce real characters behind the facts, and tell a true story with action and excitement in order to communicate information about their subject in a way more compelling than would generally be possible with straight exposition or academic argumentation; they could, in short, use their research as the basis for creative nonfiction.

"Creative" in creative nonfiction refers to the style—not the substance—of the work; it differs from fiction and scenario writing in the sense that nothing can be fabricated or imagined. Instead, it is based in careful research and observation, just like journalism or academic writing—making it well suited to relaying the conclusions of long-term scholarly research. The editors to which Carper and Schmid appealed were largely receptive to the idea of combining SIP and story: Laura Helmuth, a short, trim, outdoorsy brunette who seems perfectly matched to the *Smithsonian Magazine*, where she was a senior editor at the time (she is now the science and health editor for *Slate* magazine); Leslie Meredith, a slender, soft-spoken and conservative-mannered vice president and editor at Simon & Schuster; and Scott Hoffman, founding partner at the New York literary agency Folio Management, and, in contrast, a boisterous presence with a booming voice and biting wit, were all well aware of the growing endorsement of creative nonfiction by the publishing community and actively encouraged the genre among their colleagues.

The fourth editor, Kevin Finneran—tall, thin, and neatly bearded—needs to be immaculately professional in his blazer and tie because he represents the august U.S. National Academy of Sciences and its wonkish policy publication, *Issues in Science and Technology* (*IST*). Finneran was perhaps the most unlikely to buy into the idea that story can communicate policy to the general public and to experts and scholars. In his decade as editor of *IST*, Finneran had published four issues per year with eight to ten articles per issue, and not a single one in the creative nonfiction genre. Yet by the end of the pitch slam, he had committed *IST* to publishing pieces produced by Schmid and Carper and eleven other scholar-communicator pairs. It was up to them to figure out how to turn scholarship into story.

Lee Gutkind and Dave Guston were responsible for bringing Schmid and Carper—and the eleven other teams—together. Gutkind, characterized by *Vanity Fair* as "the Godfather behind creative nonfiction," led the field's vanguard by creating graduate programs in creative nonfiction at the University of Pittsburgh and Goucher College in Baltimore. Guston, co-director of the Consortium for Science, Policy & Outcomes (CSPO) at Arizona State University—recently ranked tenth in a list of "global go-to think tanks" in the area of science and technology—is a published and highly cited scholar in science and innovation policy who has always aimed for that broader audience. Their idea, simply enough, was to bring promising writers and scholars together, in pairs like themselves, and train them in creative nonfiction.

Gutkind and Guston had seen that while science scholars have found a popular audience, SIP scholars had not quite yet done so. Communicating science demands a mind-set and an understanding of processes and terminology that seems foreign and elusive to many readers. Yet narrative books about science, technology, and related topics have always found readers. Lewis Thomas' *The Lives of a Cell* (1974) inspired many young men and women to become doctors and scientists, while Tracy Kidder's *The Soul of a New Machine* (1981) ushered in the computer age. Not to mention Carl Sagan's *Cosmos* and *Gödel, Escher, Bach* by mathematician Douglas Hofstadter. The latter three authors each won the Pulitzer Prize in general nonfiction roughly three decades ago. Today, more books about science and technology are being published in this narrative form and represent a bright spot in the otherwise struggling publishing industry.

Communicating science policy, however, squares the complexity and challenge of communicating science because policy is as foreign in its practice to most people as science is. Yet such global challenges as the Fukushima nuclear accident or climate change or the risks of emerging technologies, as well as such domestic concerns as the innovation economy or the immigration of high-tech workers, demand attention to their social and policy aspects, not just their technical ones.

While most of us burnt a test tube or dissected a frog in chemistry or biology lab in high school, it is likely that our closest exposure to the policy process is pulling a lever in a voting booth every couple of years. And precious few of us have given any thought to the one of every eight discretionary dollars the U.S. federal government spends on scientific research and development (R&D), or the role that data and standards play in fostering environmental health and safety, or how people interact with complicated technological systems like nuclear power plants.

Compounding the problem, SIP scholars—who often have undergraduate degrees in science or engineering and graduate degrees in the social sciences—are not experienced or comfortable writing for the general public and explaining what they do. They'd rather write for policy makers, whose background and sensibilities are more like their own. Using social science research techniques like interviews and observation, many SIP scholars are often trying to distinguish their work from that of journalists and convince skeptical readers that their accounts should be deemed reliable knowledge and not "just stories." So even as doctoral and master of fine arts programs in creative nonfiction are spreading across the globe; institutes and programs in narrative science, narrative history, narrative law, and narrative medicine are blossoming; and subspecialties of narrative genetics, narrative neuroscience, and narrative psychiatry are emerging, SIP scholarship is largely stuck in its traditional mode of plodding academic articles tucked away in journals whose subscription base would complete the guest list of a medium- to large-sized wedding.

And while it is true that more and more writers and journalists are recognizing the value and excitement of writing about science, the fact is that science writers are generally more interested in reporting about the excitement of scientific discovery or the allure of emerging technologies, rather than policy, of which they may be less aware, or in some instances totally unaware. Gutkind recently approached a group of journalists, all of whom were veteran reporters and journalism professors, and about half of whom specialized in science reporting, to ask for a letter in support of a writing project. Two of the group had no idea what was meant by science policy. One veteran reporter and director of a prestigious journalism program asked, "By science policy, do you mean like what Oliver Sacks does?" He was referring to the neurosurgeon most known for his book of creative nonfiction essays, *The Man Who Mistook His Wife for a Hat.* Another, a veteran science writer who teaches for a prestigious science writing program, said after studying a science policy website that she could see how policy related to some of her colleagues who write about genetics or robotics, but she couldn't see how policy could relate to her field of astronomy.

Because of journalists' and editors' lack of awareness of SIP issues, important critical perspectives are often left out of even the most successful writing about science. A good example is Michael Specter's essay in *The New Yorker* in December 2010 about synthetic biology. Specter produced an incredibly lively, readable piece about cutting-edge biology and the scientists, like Jay Keasling at the University of California, Berkeley, who do it. Better still, he coaxed the scientists into discussing some of the ethics and policy questions that surround the field, including issues of safety and security in the creation of organisms novel to evolution (and public health). But Specter stopped with the scientists' own accounts of these policy and ethics questions—either not recognizing or not pursuing the work of talented and insightful SIP scholars whose views of what synthetic biology is like, what its prospects are, and what its risks and benefits might be differ from those of the scientists involved in the research. The result was that readers might not know that there are compelling and legitimate ways of viewing synthetic biology that do not appeal to the scientists' narrative of its inheriting the millennia-long and munificent mantle of agriculture and husbandry.

In contrast, Rebecca Skloot's *The Immortal Life of Henrietta Lacks* also combines good science with fine narrative storytelling—but simultaneously illuminates serious challenges for science policy and research ethics. Skloot received an MFA in creative nonfiction and served as an assistant editor at *Creative Nonfiction,* the first literary magazine to publish creative nonfiction exclusively. Prior to publishing her book, she also taught creative nonfiction at the University of Pittsburgh and the University of Memphis. Skloot excavated the story of Henrietta Lacks and personalized through narrative the story of how immortal cells, the HeLa line, taken from her ultimately fatal cervical cancer, became vital to the development of the polio vaccine, as well as drugs for treating herpes, leukemia, influenza, hemophilia, and Parkinson's disease; how her cells helped uncover the secrets of cancer and the effects of the atom bomb, and led to important advances like cloning, in vitro fertilization, and gene mapping; how, since 2001 alone, five Nobel Prizes have been awarded for research involving HeLa cells; and how her family, the descendants of slaves, were unaware of these contributions for decades after her death. In creating her *New York Times* bestseller, Skloot also drew significantly on the work of SIP scholar Hannah Landecker.

So, to trawl for the next Skloot and encourage the next Specter, Gutkind and Guston conceived "To Think, To Write, To Publish." Through broadly circulated calls, they recruited a dozen early career SIP scholars (of more than three dozen applicants). They included a philosopher of science concerned with the narrow scope of bioethics in the U.S., a sociologist who investigates the interaction of science and the marketplace, and an expert in marine science and policy who studies how small fisheries innovate to prevent bycatch of protected species. They recruited a similar dozen early career communicators (from more than 240 applicants!), including a recent MFA graduate in creative nonfiction interning at *Wired*, a children's literature specialist and a contributing editor to Otata.org, a Chicago-based collective of photographers and writers, and a publisher of genre fiction from Portland, Oregon. The 12 collaborative two-person teams, composed of a "next generation" SIP scholar and a "next generation" writer, would learn creative nonfiction and narrative techniques and would write a creative nonfiction essay together, drawing on the scholar's research.

To Write

In which two actual people survive an exasperating collaboration to complete a task neither could have done alone.

Six months after the pitch slam, the second of the "To Think, To Write, To Publish" pieces for *Issues in Science and Technology* arrived in the inboxes of the organizers and the editor, Finneran. To Gutkind, it left something to be desired: "What we're looking for," he explained over the phone to authors Gwen Ottinger and Rachel Zurer, "is something with more... *narrative.*" Gutkind went on to point out several places where their collaborative piece about Ottinger's research had lapsed into lifeless exposition.

When Zurer called an exasperated Ottinger to regroup, she honed in on a particularly abstract part of the article. Ottinger—a Berkeley trained anthropologist with an undergraduate engineering degree and knack for asking the kind of question informed by both parts of her training—had spent most of a decade studying efforts to measure ambient air concentrations of toxic chemicals in communities next to oil refineries and other petrochemical facilities. She wanted their piece to make the point that more extensive air monitoring was only useful if it was combined with research on health symptoms in the community. The argument hinged on a criticism of existing regulatory standards and screening levels for toxins in the ambient air, which serve as the yardstick against which we can measure air quality: they vary greatly depending on who is setting them, in large part because they are—almost necessarily—one part research, one part extrapolation, and one part expert judgment.

The manuscript they submitted had *said* as much. Zurer, a newly minted MFA deeply committed to her craft of creative nonfiction and gently dogged in her desire to understand the subjects she wrote about, now wanted them to figure out how to *show* it. She suggested making standard-setting into a scene in their narrative. But a regulator sitting at a desk and sifting through a stack of scientific studies, deciding what safety factors to apply and how to combine the results, hardly seemed cinematic. Besides, Ottinger complained, if they made a regulator their main character, they could only show the process of making one set of standards, and not the larger context, where uncertainties, disparities, and omissions became obvious and limited the usefulness of standards to residents and policy makers trying to understand whether emissions from an industrial facility were affecting a community's health.

"So who is seeing that larger context? Who is making the argument?" Zurer asked, perhaps hoping that some charismatic community activist could become the cornerstone of their article.

"Nobody!" Ottinger snapped. Then she sighed, "I mean… I am."

Ottinger went on to explain that this was why it was so important to her to publish her research in a policy journal like *Issues in Science and Technology*. In general, scholars, especially those early in their careers, confined themselves to academic journals, where every article is reviewed by two to four experts on the subject matter before it can be published. The peer review process certifies the quality of one's research, and despite such journals' small readership, publishing in them establishes an academic's reputation and allows her to advance in the university. Writing a narrative piece for *IST* would hardly help Ottinger's tenure case. Yet she felt she had something to say that people beyond her hundred closest colleagues needed to hear.

Community and activist groups she studied had been advocating for—in some cases quite successfully—more ambient air monitoring on petrochemical facility fence lines. But the monitoring data would be of little use without better tools for interpreting it. And building those tools would require that the problem be acknowledged, not just by community groups, but also by scientists and policy makers who could offer resources to the project.

It was no surprise that Ottinger would identify problems through her research that others hadn't recognized, or that she would take policy positions that were not necessarily advocated by the activists whose campaigns she studied. It is, in fact, part of the nature of scholarship. Academic researchers have a unique perspective: they can take a big-picture view, looking at large structures as well as particular circumstances, and they can take years to untangle what is happening at multiple levels and see how it all fits together. Policy recommendations based in scholarship are thus likely to be far-reaching, long term, and relatively novel.

But creating a narrative that would not only convey but also truly drive the communication of a scholar's perspective on air monitoring in communities at the fence line challenged Zurer and Ottinger. Good stories have characters, they have conflict, and they have resolutions. What the scholar-communicator pair had was an ongoing problem that no one was even fighting about yet—except maybe for Ottinger herself, and then only through her academic writing and occasional advice to activist groups.

"So we should do that—tell your story," Zurer said after listening to Ottinger explain how she had come to the conclusions that she wanted their article to convey.

Zurer's suggestion was a radical one: policy scholars write about their observations and analyses, not about themselves and their learning processes. Nor do journalists allow themselves to become characters in their own stories. But having tried everything else they could think of, Ottinger squelched her discomfort and began, "I was at the most undignified moment of moving into my new office—barefoot and on tiptoes on my desk, arranging books on a high shelf—when one of my fellow professors at the University of Washington Bothell walked in to introduce himself."

A story finally emerged. Within the first two paragraphs, the unnamed colleague provides the conflict the narrative needed, asking the vexed question of whether emissions from oil refineries really make residents of the communities at their fencelines sick. Ottinger's struggle to answer "The Question" then sweeps the reader with her to an Oakland, California internship, to her fieldwork in Norco, Louisiana, to Zurer's meeting with activist Marilyn Bardet in Benicia, California, and then back to Ottinger's office and, finally, inside her head where she formulates a plan more extensive environmental health monitoring upon which better ambient air standards could be based, by imagining the answers that *could* come from extensive new air monitoring programs in communities like Benicia:

> *I wandered off to the faculty holiday party conjuring a new daydream: The National Institute of Environmental Health Sciences would call for proposals for studies correlating air monitoring with environmental health monitoring; the EPA, making ambient air toxics standards a new priority, would demand that data from fenceline communities be a cornerstone of the process; and Marilyn Bardet would seize on the new opportunities and make her community part of creating a better answer to The Question.*

Interweaving policy recommendations with Ottinger's quest for answers earned the approval of both Finneran and Gutkind; Ottinger and Zurer's article became part of *IST* readers' introduction to science policy in narrative form.

To Publish

> *In which at least some of our scholars and writers, having overcome the trials and tribulations of thinking and writing, succeed in their ultimate goal.*

Prior to the pitch slam, the agent Scott Hoffman and the editors Laura Helmuth, Leslie Meredith, and Kevin Finneran had offered their advice to the newly forged narrative-SIP teams. But at times they didn't seem to get it. To be sure they were offering helpful advice. But their experience seemed to be too often editing material in which science was the subject, and not the object, of study. In which scientists were the narrators or central protagonists, and not problematic characters in need of explanation themselves. In which the book sells because it is selling the excitement of science. In which one-word titles and Nobel laureates demand attention to their own view of the future. While this is a necessary evil in today's challenging publishing environment, there were other important considerations.

"How much room is there in commercial nonfiction for ambiguity, ambivalence, and criticism around science and technology, rather than promotion and adulation?" Guston asked. A general laugh followed the response, "That's a tough market!"

Helmuth from the *Smithsonian* took the idea of criticism in the wrong direction, responding that "It's hard to do a book-length treatment of scientists misbehaving, although people have done it."

But critical doesn't always mean searching out the seamy side of science—the way, for example, journalists William Broad and Nicholas Wade did in *Betrayers of the Truth,* a book about scientific fraud and misconduct drawn from their front-page *New York Times* stories. Rather, SIP scholars approach science with a level of constructive, learned, critical engagement the way a food critic approaches a restaurant or a movie critic approaches a film. A Roger Ebert for science and innovation would understand not only the role of the actors who perform the science, but also the research directors who organize it, the producers who finance it, and the audiences who are moved by it. He would understand if the purpose is profit, passion, politics, or pure creativity. He would leave room for ambiguity and ambivalence beside the adulation.

Hoffman, the agent, was modestly more expansive in his understanding of criticism, but nevertheless maintained that readers "want to know what is possible rather than what isn't possible. And so what you're selling is to a certain extent a fantasy or a possibility." But properly done, creative nonfiction derived from SIP scholarship can extend possibilities rather than rein them in. Just like in art class, the criticism goes hand-in-hand with appreciation. SIP scholars recognize science as a cultural achievement worthy of praise, and they acknowledge effective, legitimate science and innovation policy as doubly remarkable in its success. The desire to examine and account for science and innovation doesn't diminish that appreciation; indeed, it stems from it. Collaborative story-telling between scholar and writer both enables a critical perspective and allows scholars' wonder to shine through by adding color, richness, and texture to central analytical points. The combination of perspective and richness make it easier for a reader not immersed in the worlds of science or science policy to nonetheless appreciate them as part of civic life and not pristine, isolated, and remote from it.

The editor Meredith pointed toward ambivalence and even controversy as a challenge to editors and writers alike, but also held it as a potentially productive force. After learning the hard way from editing a book that was a "nice story" but was universally reviewed as presenting a one-sided view of the science, Meredith said that she now asks her expert authors, "'Who dissents from you in all this?' And it provides narrative tension if you have to defend yourself." But she still saw dissent in terms of a controversy within the scientific community, rather than a conversation among scientists and non-scientists about the purpose and meaning of the enterprise and how scientific practice reflected those aims. Meredith, like Michael Specter, hadn't yet included SIP scholars—as credentialed, hard-working, public-regarding, and brilliant as the scientists themselves—who want to engage in this critical dialogue with scientists and facilitate such a dialogue with the public.

The thinking and the writing have shown the profound challenge that communicating science and innovation policy faces. But it was harder still, the participants were to discover, when put into practice. Several of the collaborations foundered on all-too-familiar shoals of conflicting professional demands and incompatible agendas between scholars and communicators. Some modes of failure were perhaps induced by the organizers. For example, because of other demands on the group of scholars, they were not tutored in creative nonfiction together with the writers, leaving them to learn a new craft from their partners, who were in some cases fresh out of MFA programs and still working to master it themselves. Further, the structure of professional training meant that "young" scholars were as much as a decade older than their "young" communicator counterparts. The unions of writers and scholars were also arranged by Gutkind and Guston, leaving each pair to struggle to discern the meaning of their match.

Yet through the completed work, Finneran saw how the genre could make the SIP issues that his journal deals with more accessible and appealing. As he wrote in his "New Voices, New Approaches" introduction to the series of articles in *IST*, "[T]his is hard to believe [but] some people would rather read a compelling story than a meticulously organized piece of rigorous academic argument." As much as even the published essays bear some scars of difficult collaborations in thinking and writing, as Finneran says, "the analysis is as perceptive as the story is engaging." They also show that creative nonfiction is, ultimately, a strong genre for what we in science and innovation policy want to say.

For Meera Lee Sethi and Adam Briggle, "Making Stories Visible" performs for synthetic biology what Michael Specter's piece did not: it shows how scientists function in stories, and potentially controversial ones, at that. Detailing the personal narrative of one true-to-life character—from comic book- and science kit-loving kid to high profile SIP analyst and master furniture maker—Sethi and Briggle highlight the variety of ways we can tell stories about synthetic biology and thus the variety of policies we might have for governing it. They also wield a metaphor, the Geiger counter built by their lead character as a child, to remind us that both science and story are built from powerful but elusive elements that take technique and patience to reveal.

For Sarah Whelchel and Elizabeth Popp Berman, "Paying for Perennialism" exposes how human lives and futures are tied up in complex systems of money, policies, nature, knowledge, and technologies. Perennial grain crops could be a great boon for the environment, reducing the stress on soil that annual plantings and regular tillage induce and increasing through larger root systems the amount of carbon that such plantings would sequester. Whelchel and Popp Berman weave together the narratives of three men—a Kansan agricultural activist, a Washington State wheat breeder, and a Cornell plant geneticist—to show how at least some in the agricultural community are attempting to perform long-term, high-payoff research on perennials when they are opposed by declining federal research budgets, hostile corporate agendas, and a skeptical and even complacent scientific community. Yet their stories, colorful and moving, may fade to irrelevance without policy makers to bestow priority on perennials research.

For Sonja Schmid and Ross Carper, "The Little Reactor that Could?" demonstrated that you can't tell the whole story of science and innovation without both people and things as actors, and without the relationships between them being part of the central dynamic. These dynamics are often acknowledged only metaphorically, and—like good creative writing—science and innovation policy can be ruled by a dominant metaphor: Schmid's Russian informant likened supposedly identical, large nuclear reactors to "children; each one is different." Schmid and Carper ask quite reasonably of small, modular nuclear reactors if their metaphoric standardization as "batteries" can reliably eliminate the need for humans to be responsible to the subtleties and complexities of individuality.

For Roberta Chevrette and Angela Records, "Living and Breathing Plants" vividly dramatizes the disconnect between researchers and policymakers—making the argument that scientists need to be involved in the policymaking community—a necessity for national health and agricultural prosperity.

And for Gwen Ottinger and Rachel Zurer, the narrative in "Drowning in Data" helped show that unknowns in science—"The Question" of whether emissions from refineries are really making people sick—are as important as the knowns. In order to expose this problem of the ambivalence of data, they had to break with conventions of both academic writing and creative nonfiction by making an author the story's main character.

In response to Guston's pre-pitch slam question, Finneran had agreed with his fellow editors that publishing creative nonfiction around science and innovation is "a challenge, it's an extraordinarily difficult challenge," especially "without becoming seen as 'anti-science.'" But he hoped that the people involved in this project "can start us down this path." Finneran declared that we would all like to see "a more ambiguous, more richly thought-out critique and understanding of science, so that it isn't so remote and so distant from ordinary people.... You get it right, we'll publish it." The writer-scholar teams thought and wrote. Finneran kept his word, and they published.

Now, several years later, the results of this unique experimental program are quite promising. *IST* has published many of the collaborator's essays. One of the communicators has launched an online creative nonfiction social action journal. Another "next gen" participant has become publisher of a medical science-oriented book series. Three of the collaborators have been offered speaking or writing engagements as a direct result of their efforts. And a follow-up program, with a much more comprehensive plan, has been funded by NSF. The ambiguous, nuanced, and narrative critique of science that Finneran envisioned stands poised to become a force in public discourse about both science and science policy.

2

MAKING STORIES VISIBLE

Adam Briggle and Meera Lee Sethi

A little before lunchtime on December 6, 1957, when the United States made its first attempt to match the triumph of Russia's *Sputnik 1* by launching its own *Vanguard TV3* satellite into orbit around Earth, David Rejeski was one of millions of wide-eyed American grade-schoolers, raised on a steady diet of science fiction stories, whose day was being thrillingly interrupted by the future. Chins propped against crossed arms, ears held closely to an array of radio speakers perched atop desktops—the glory of being allowed to bring their own radios to school, of all places, and actually take them out in the middle of class!—David and his classmates listened breathlessly to the broadcast. The disembodied voice of the announcer from Cape Canaveral trembled through the static; the strange, stormy rumble that marked the sound of liftoff spread across the airwaves. Unhappily, the *Vanguard* was an ill-fated rocket that never made it into space; it lost thrust just two seconds after blasting into the air, sinking back onto the launch pad like a bad firecracker and disappearing into the flames that exploded from its still brimming fuel tanks. The new year would have to come and go before the *Explorer 1* finally became the first successfully launched U.S. satellite and the nation's next dramatic move in the Cold War space race.

But for at least one little boy—already spending every spare moment building rockets and operating ham radios, already enshrined as the proud president of a basement chemistry club whose activities scared his mother half to death, already deep into a self-administered program of scientific study spanning the fields of aerodynamics, electronics, and physical matter—there could have been no greater rush than the one that came from huddling around the radio in school that day. It wasn't, you see, just the news he was listening to. Eavesdropping on the fate of that slim 72-foot rocket plucked David straight out of the four bland walls of his classroom and plunged him into a sensationally exciting issue of one of his beloved Captain Marvel comic books. For when science was a tool caught between good and evil (as it seemed to be that morning, and as it seemed the day a few years later when David figured out how to put together a homemade Geiger counter so he'd know if it was safe to walk out of his house after the Bomb fell), the stories, books, and movies that filled the young boy's imagination became a powerful way to understand how science functioned in the real world.

More than five decades later, David Rejeski has grown basketball-player tall and cultivated a rumpled shock of salt-and-pepper hair that brushes over his ears; together with a matching mustache, it gives him a little of the look of a leggy Einstein. He has big, graceful hands that he still doesn't shy away from getting dirty. The first degree Rejeski earned was a B.F.A., and in one of his lives he dreams up and sculpts beautiful pieces of handcrafted furniture, like smooth hardwood tables whose tapering legs are inspired by the shape of chopsticks or whose surfaces bear the intricate texture of thousands of individually chiseled facets. In his work life, though, the one in which he finds himself wearing the uniform of suits and ties and uses those wood-calloused hands to gesture with broadly as he speaks before government officials, Rejeski grew up to be a scholar of science, policy, and technology. Among many other responsibilities, his current job as the director of the Woodrow Wilson Center's Program for Science and Technology Innovation involves assessing how the public understands both the promise and the peril of emerging scientific endeavors. He studies, in other words, fields such as nanotechnology and synthetic biology that mark today's bold new frontiers in science the way space travel did 50 years ago. And though it might seem surprising, he's still got Captain Marvel on the brain.

These days, though, when Rejeski encounters stories about science, he's a little less wide-eyed and much more savvy about where they come from and what they mean. He was paying close attention, for instance, on the bright May day this year when biologist J. Craig Venter announced his institute's historic accomplishment: creating the first viable bacterial cell whose genetic material had been written in digital code and then synthesized in a lab. Venter was standing at a press conference podium wearing a sober blue jacket and shirt, not sitting cross-legged around a campfire with shadows at his back. His voice was nonchalant, even matter-of-fact, not theatrical. But Rejeski could see that Venter, who has a reputation for being a renegade researcher with little regard for the "rules" society attempts to place on science, was telling a powerful story designed to downplay the potential risks of synthetic biology. Previously, Venter had compared the process of working with genome base pairs to solving a jigsaw puzzle or connecting the spools and sticks of Tinker Toy pieces—the kind of analogy that has always irritated Rejeski mightily. "When you use those metaphors," he grumbles, "when you talk about building blocks and Legos, it infantilizes the science. It becomes something that children can play with, and therefore it can't be dangerous." At Venter's press conference, that storyline grew up a little; but it was still calculated to simplify what is an incredibly complex biological process.

"This is the first self-replicating species we've had on the planet whose parent is a computer," Venter said to a roomful of journalists. He spoke of long months spent "debugging" errors in the synthetic DNA and of "booting up" the cell into which it had been transplanted. Finally, he explained that the scientists who had created the cell, known as Mycoplasma mycoides JCVI-syn1.0, had "encoded" a series of messages into its genetic material, including the names of authors and key contributors, the URL of a Web site, and three literary quotations about the nature of discovery and creation.

By framing his work through the narrative of computer engineering, Venter was crafting a story about synthetic biology that presented it as a safe, repeatable, and controllable technology. Life, ran the story beneath his words, is essentially information. Organisms are information-processing machines. Creating life is like making a machine; if its design contains errors, we will find and fix them. And like a machine, the nature of a synthetic organism is so malleable to engineering that its DNA can be stamped with its creators' intentions. "What Venter's doing," says Rejeski, "is making use of an engineering narrative that sends a message to the policy people and the public that all this has a high degree of controllability. People tend to think, well, engineers do a fairly good job. Most of the time, bridges don't fall down. But a cell is essentially a stochastic system, and we don't have that kind of control over it. Venter's got enough of a microbiology background to know better. He's using a reassuring story that makes everything seem much simpler and less risky than it really is."

Science and storytelling appear antithetical. Science deals in a non-narrative form of rationality, offering facts where stories offer interpretations. But Rejeski pushes back on that easy dichotomy. "Storytelling and narrative are absolutely critical to science," he will tell you. "The public uses stories to understand science, and so do scientists, whether they're doing it on purpose or not." One place where the two realms intermingle is the space Rejeski happens to inhabit every day: evaluating the human significance of new scientific discoveries. What is life? What would it mean to live in a world where humans synthesize life?

Lacking a single "objective" answer to these questions, our responses to them depend on framing and perspective—aspects of storytelling. The philosopher Fern Wickson made this clear when she closely examined nine common cultural and scientific narratives about nanotechnology, each "a story that begins with particular presuppositions and ends in support for particular areas of nanotechnology development." In some, nanotechnology is shown as controlling nature; in others, transgressing its boundaries or treating its ills. Yet though these stories are clearly distinct, Wickson writes, each is presented "as a simple description of the way things are... this often masks the beliefs that underlie each of the different narratives and the research directions in which they tend to lead." In other words, many narratives about science are invisible. Not recognized as stories built on particular assumptions and expressing particular points of view, they can seem to be simple accounts of reality. This is particularly true of stories that accumulate around emerging disciplines such as nanotechnology and synthetic biology, whose applications, implications, and limitations are not yet well understood by the public or scientists themselves.

Rejeski is among the few working in science who have made this issue his business, a fact that he laughingly admits can make him feel a little like Pandora: constantly opening the lid of a box most researchers and policymakers would rather keep shut tight. Yet with questions of law—Should we press on with this technology? With what limitations?—answers depend on the story one chooses. It is important to make those choices with our eyes open to the ways different stories, including those told by scientists and engineers, frame and interpret reality. This is the point Rejeski emphasized in July 2010 during his invited testimony before the newly formed Presidential Commission for the Study of Bioethical Issues (PCSBI). Chaired by University of Pennsylvania president Amy Gutmann, the PCSBI is the government body that was charged by President Obama with assessing the risks and benefits of synthetic biology as soon as Venter's feat went public.

At about 9 a.m. on July 9, Rejeski, dignified in a dark gray suit that hung just a hair too large on his shoulders and a striped tie that he reached up to smooth several times as he began to speak, took the place assigned to him in the cool carpeted conference room of the Ritz-Carlton hotel in downtown Washington, DC, where the PCSBI had chosen to hold its first round of meetings. To his front and sides were the 13 members of the commission and two fellow panelists; together, these central attendees were seated at tables that formed a closed square. Behind them, and out of Rejeski's sight, about a half dozen rows of chairs were slowly filling up with members of the public. He didn't need a good view to know that these probably weren't teachers or electricians or firemen who just happened to have a personal fascination with genetics; instead, the audience was made up of a small and very specific set of people with a vested interest (money, mostly) in synthetic biology. Industry insiders. In fact, the meeting, which began with Gutmann introducing a designated federal officer to "make it legal," resembled nothing so much as the formal gathering of a board. Which is perhaps why it was so much fun for Rejeski to know that besides graphs and other data, in a few minutes he was about to show these people slides of comic books, movie posters, video games, and cartoons.

He was the first speaker of the day, and he began simply enough. "Let me start," he opened, "by saying that we have devoted about six years of our time... trying to bring the voice, or voices, of the public into the conversation about science policy on emerging technologies." If you weren't paying attention, you might have missed his next sentence, delivered almost as a throwaway as he searched on the table for the clicker he would need to control the rest of his presentation. It didn't draw a laugh from the crowd but was obviously charged with a deeply dry humor that emerged from Rejeski's sense of how little attention is paid to this kind of work. "In terms of how we do this?" he said, "It's pretty easy: We talk to them."

In the past few years, he'd traveled from Spokane to Dallas to Baltimore, Rejeski said, simply asking people what they knew and how they felt about synthetic biology. And what he'd found was that because most people don't understand the science behind it, the combination of these two words tends to set off a fast-moving train of loose associations in people's minds, fueled by half-remembered news stories. "The train," he explained, "goes something like this: Synthetic biology, is that like artificial life? Is that cloning?" Rejeski's pace, normally measured and thoughtful, became brisker as he counted out the links, which he said took most people about 15 seconds to get through. "Is that stem cells? Is that GMOs?" He stopped. Raised a pair of bushy eyebrows. When asked about the possibility of someone creating synthetic life, Rejeski explained, there was a clear trend among the people he met: "'I'm worried about this.' Over half. 'I'm excited about it.' Less than half." But if they didn't know much about this field of science, why exactly would public perception skew toward fear? Though most people in the room wouldn't realize it, Rejeski's answer would take him back to the little boy he'd once been.

For most of the previous ten minutes, the images appearing on Rejeski's slides had been perfectly conventional. True, a Gary Larson cartoon had sneaked in that gave him great pleasure to include. (In the first panel of the original, a man admonishes his dog to stay out of the garbage; in the second, a word-balloon shows the only thing getting through to the dog: its name. Rejeski had searched the Internet for hours to find the cartoon, one of his favorites, and carefully modified it to show how the public understands scientific communications about synthetic biology. In his version, the balloon in the top panel read "synthetic bacterial cell genome… artificial DNA base pairs… sustain life replicating;" the one on the bottom, "blah blah SYNTHETIC blah blah blah LIFE blah." When it went up, a smile came into view under the speaker's mustache that he couldn't quite hold back.) But otherwise, Rejeski's testimony had largely been illustrated with a series of neat color-coded bar graphs depicting the vast amounts of data he'd collected. He'd "stuck to the script," as he would later put it.

In the last minute of his testimony, though, his tone shifted. He returned, for just a moment, to the way in which he'd first begun to relate to science: through the lens of story. "Human beings," said Rejeski, quoting the late novelist David Foster Wallace, "are narrative animals. That is how we understand science." Even as he spoke, he clicked over to one of his last and most surprising slides—one that hadn't even made it into the first version of his testimony, but that in the end he couldn't resist using. It was full of stories. "This," Rejeski began cheerfully, pointing to a colorful vintage comic book cover complete with costumed superhero jetting across the sky, "is *Captain Marvel and the Wonderful World of Mister Atom*." He gestured to the right, where he'd placed images from *Spider-*

man 2—the looming villain Doc Ock standing with his back to us, four long metallic tentacles twisting out of his back like snakes—and a frightening screenshot from an Xbox 360 game called *Bioshock,* set in a post-apocalyptic world populated by insanely violent, genetically mutated humans. Further down, if you'd been in the Ritz Carlton that day, you'd have seen the cover of Michael Crichton's bestselling novel *Prey*: a black, buzzing cloud of tiny escaped nanobots darkening the sky like a Biblical plague; and an image from the new genetic engineering horror flick *Splice*: a bald, hoofed, three-fingered humanoid with huge blank eyes and a perky tail, crawling on top of a lab table.

"These are deep, deep narratives," Rejeski said. He described large circles in front of him with his hands as he talked, as if pushing the stories towards the commission members, willing them to understand their importance. Rejeski himself felt that importance keenly. These stories, he knew, were the primary source of the unease he'd sensed about synthetic biology from the people he'd talked to; these stories, functioning mostly on an unconscious level, were the fire fueling fears about escaped organisms, new terrorist threats, and the hubris of designing life. "The thing that the scientists have to understand," Rejeski concluded, in a voice that could not have been more urgent and sincere, is that "people will fall back on these narratives long before they will ever pick up a biology book. And they are incredibly pervasive, ubiquitous, and powerful."

When he thinks about it later, Rejeski still isn't sure how his testimony was received. He was warmly thanked by several commission members, he says, but can't tell whether those were simply formalities. Frankly, he says, he's just glad nobody attacked him in the corridor or called him crazy to spend so much time talking about comic books and movies. "I guess that means I'm still kind of tolerated," he chuckles. One thing Rejeski does admit is that the stories he chose to deconstruct in his talk aren't the only narratives about synthetic biology that have an impact on the discourse; not by any means. As his own frustration with Venter's conveniently adopted metaphors indicates, scientists themselves are not immune from the storytelling impulse. In fact, the day before Rejeski spoke, PCSBI heard from an array of scientists and engineers whose narratives, unlike those of Captain Marvel and Michael Crichton, remained largely unexamined—invisible to the substance of the debate.

One of these scientist-storytellers was Drew Endy, assistant professor of bioengineering at Stanford University and the director of BIOFAB, a facility that makes standardized DNA parts freely available to academic labs, biotechnology companies, and individual researchers. If Rejeski is approachable and avuncular, Endy, whom a recent Stanford Magazine profile called synthetic biology's "most fervent evangelist" and described as emitting "a sense of barely contained energy," has the charmingly intense air of a round-spectacled John Lennon after a recent haircut.

Endy, like Rejeski, is well aware of how much more powerful scientific narratives become when they are interwoven with popular culture. In 2005, *Nature* published a comic book written by Endy titled *Adventures in Synthetic Biology*. In its 12 brightly colored pages, Sally the Professor instructs "Dude," a plucky young science student, about the basics of synthetic biology. Dude's mastery of the subject comes from experimenting with a bacterium with the friendly name of "Buddy." Through his efforts, Dude learns that the genome is the "master program" and that organisms can be "reprogrammed" to perform unprecedented functions. The story is suffused with a sense of adventure and, yes, scientific playfulness. Life is portrayed not only as infinitely malleable, but also as essentially interchangeable with human artifacts. After all, life is the "stuff" Dude is "building," and he does so with inverter devices that incorporate bits of DNA. The story does contain one accident in which Buddy explodes, but this happens early on and Dude learns from his mistake. The wildly optimistic, even hubristic, message of the comic is that with sufficient knowledge humans can master life and reprogram it to suit their desires. Its last lines, which could have come straight out of a Dick and Jane picture book, read "Look at us! We're building stuff!"

Not surprisingly, where Rejeski drew the commission's attention to the dystopian stories of pop culture, Endy focused on the utopian potentials of synthetic biology. He did so through a subtle storyline that drew on a well-established analogy between the genetic code and the structure of human language. And in so doing, Rejeski later reflected, he was making a "brilliant" narrative move that tied synthetic biology to an old, unthreatening, and much-loved technology.

Dating back to James D. Watson and Francis Crick's own descriptions of the structure of DNA, the linguistic metaphor for understanding the genome refers to the chemical bases that make up each molecule of DNA as "letters." As such, each three-letter codon, or unit of genetic code, becomes a "word," and the genome itself is the ultimate publication: "the book of life." And if, as Endy suggested that day in his testimony, organisms are information that can be sequenced, stored in a database, and edited, then it's easy to see the tools of synthetic biology as tools for reading, writing, and publishing. To bring this story home, Endy made use of the narrative of literature and the printing press. Today's genetic engineering projects, he pointed out, are limited to using fewer than 20,000 base pairs of DNA. "20,000 characters," Endy mused. "That gets you things like the Gettysburg Address, which is around 1,500 characters. It gets you an editorial in the *New York Times*." He nodded as he spoke, as if cementing the comparison. But advancements in the tools belonging to the field of synthetic biology promised a future in which genetic engineering could involve a 400-fold increase in the number of characters (the number of DNA base pairs) that scientists could put together. "What," Endy asked carefully, in a rhetorical move worthy of Socrates, "would be the sort of stuff you could write with 8 million characters? You certainly get one-act plays, like *No Exit*. You get *The Color Purple*, which is not even a million characters. You even get *War and Peace*."

As with his comic book, Endy's testimony framed synthetic biology as a creative activity with limitless possibilities. But in order for such inspired human creativity to truly flourish, Endy emphasized that it was imperative for government policies to be instituted that would facilitate what he called "freedom of the DNA press." For instance, more public funds should be channeled into synthetic biology research, and individuals should be as free as possible to use this technology. "The ability to synthesize DNA in genomes is like a printing press," Endy explained, "but it's for the material that encodes much of life. If one publisher controlled all the presses, that would give a publisher tremendous leverage over what's said."

There could hardly be a more seductive narrative about synthetic biology. Writing an organism's DNA, ran Endy's hidden story, is fundamentally a creative endeavor. It is a process by which we might reach the greatest heights of artistry and express the most profound truths, as long as our efforts are not stifled or censored. Like freedom of expression, the freedom to create new forms of life should be a fundamental right. Who knows where the next Shakespeare or Melville will come from? To restrict access to the tools of synthetic biology would be a form of censorship.

Are these valid assumptions and an appropriate framing of synthetic biology? Perhaps. But perhaps not. Unlike human languages, for example, the "alphabet" of DNA does not lie inert on a printed page but takes physical form when the proteins it encodes are synthesized. And although it is easy to accept that the exercise of artistic creativity demands little or no government oversight, it is less clear that the same is true of all scientific explorations.

So why, despite its flaws, did Endy choose to frame his discussion of synthetic biology inside this particular narrative? Rejeski has an idea. What Endy was telling the commission, Rejeski points out, "is a story about scientific evolution, not revolution. It says that this is just an extension of existing science, and there's nothing disruptive or novel about it. Remember people doing work on recombinant DNA in the 1970s? They said the same thing. Oh, we're just doing what nature's been doing for a long while. It was a convenient story. And in that sense, Endy was brilliant to pull it back to the Gutenberg printing press. I mean, who's afraid of the printing press?"

Tellingly, however, whereas the cultural narratives raised by Rejeski the following day were immediately dismissed as overblown, no one in attendance on this occasion—not commission members, other speakers, or anyone in the audience—approached the narrative framework that lay beneath Endy's words with a critical eye. No one wondered whether this storyline might not be, in its own way, just as mythic as the Trojan horse or Pandora's box. Instead, the narrative remained implicit, and therefore unexamined. As if to prove the effectiveness of the scientist as storyteller, commission member Barbara F. Atkinson, then the executive vice chancellor of the University of Kansas Medical Center, raised Endy's evocation of the freedom of the DNA press during the question session that followed his panel. She had been "caught by" this comparison, Atkinson said. Could the panel members recommend specific policy recommendations PCSBI might make to support the workings of the genetic free press?

The absence of criticism directed at Endy's narrative stems from the assumption that scientists and engineers are what biologist/philosopher Donna Haraway calls "modest witnesses." They are ventriloquists for the objective world, adding nothing of their own voices. Their "narratives have magical power, they lose all trace of their histories as stories... as contestable representations, or as constructed documents in their potent capacity to define the facts. The narratives become clear mirrors, fully magical mirrors, without once appealing to the transcendental or the magical."

In opening the commission's first meeting, Gutmann noted that "it is key for this commission to be an inclusive and deliberative body, encouraging the exchange of well-reasoned perspectives with the goal of making recommendations that will serve the public well and will serve the public good." In support of that mission, the commission would go on to hear hours of testimony from engineers, biologists, theologians, philosophers, social scientists, bioethicists, lawyers, and others. It would be told by some that Venter's work is nothing but an incremental step in a long history of genetic manipulation, and assured by others that the achievement represents a complete scientific game-changer. Commission members would be urged by some to advise a near-moratorium on synthetic biology in order to prevent an unjust bioeconomy, and encouraged by others to step hard on the accelerator to bring new products to market. To frame these diverse and often conflicting views, each speaker would bring a story or stories to the table.

Crucially, however, the testimony PCSBI has heard in the months since it was first formed has not treated all narratives with equal scrutiny. Thanks in part to Rejeski's efforts, the commission has made progress in rendering visible the most pervasive cultural narratives about artificial life, seeing these stories as imperfect and unscientific constructions by artists, the media, and other mythmakers. But those who have testified before the commission have been far less likely to turn a critical eye on scientific and engineering narratives, instead allowing these stories to remain implicit and therefore invisible. To produce a truly thoughtful and deliberative report on both the practical and ethical implications of synthetic biology, PCSBI must ensure that no story, no matter its provenance, goes unexamined. It must render each of the stories it is being told about this science more visible, exposing their interpretive frames and subjecting their assumptions to critical scrutiny. Because policymaking has to happen on the basis of one story or another, it is best to inform those decisions with an explicit account of the available options. Advisory bodies must not allow a single narrative to become the invisible lens through which the issue at hand is viewed. To do so constitutes a technocratic overreaching of expert advice, because it makes one story seem to be simply the facts. Policymaking would be constrained in advance to choices within a single narrative, and policy discourse would be limited to the terms and goals established by that story alone. Advisory bodies should clarify and expand, not limit, our choices.

One welcome critical treatment of narratives was at work in the testimony of Randy Rettberg, a principal research engineer in Biological Engineering and Computer Science and Artificial Intelligence at MIT and the director of MIT's Registry of Standard Biological Parts. Speaking at PCSBI's September 14 meeting, Rettberg told a story from his youth. "When I was a junior in high school," he began, "I decided that my father was an architect of buildings, and I wanted to be an architect of computers." At the beginning of his career, he went on, that seemed an impossible goal, but thanks to a development in the field that enabled processors to be built out of a set of tiny standardized parts known as transistor-transistor logic, the dream became achievable.

The testimony that followed was striking. Having told a story with obvious connections to synthetic biology, Rettberg immediately went on to clarify the underlying assumptions behind the narrative, pointing out which ones ought to be accepted and which discarded. What was accurate about his story, according to Rettberg, is the idea that making simple interchangeable parts freely available to a large population of researchers might revolutionize the genetic engineering industry in terms of what it is able to produce and who is able to produce it. Less accurate, but much more quickly grasped by listeners, is the idea that synthetic cells are actually like little computers, with each internal component operating in a fundamentally logical manner. This, cautioned Rettberg, was "not really right."

As Rettberg's testimony illustrates, stories about synthetic biology are often based on faulty comparisons. Yet it is clear from the meetings that PCSBI has held so far that they are also a ubiquitous part of the debate, because they serve as a way to make sense of complex and sometimes contradictory scientific information. The commission should acknowledge these multiple narratives, confront them with the careful attention Rettberg gave his own story, and explicitly probe them for valid and invalid assumptions. In its report, PCSBI should outline more than one set of policy options, bolstering each one with clear justifications for its premises and articulating, where appropriate, when a proposal stems from a particular narrative about this new technology. If, for instance, PCSBI were to adopt Endy's recommendation that a substantive public investment be made in the tools of synthetic biology, it should not do so without first thoroughly examining the suppositions behind the narrative of the "DNA free press." In so doing, it will multiply and clarify options for policymakers rather than handing them just another story with its black box of assumptions.

When Rejeski was invited to speak before the commission, he thought long and hard about what he would say. Was he really going to show up for its first meeting with slides of comic books, movie posters, video games, and cartoons under his arm? How exactly would his listeners respond? "I don't think 90% of people spend a lot of time," Rejeski muses, "asking whether the narratives people tell about science are valid, or just being used as a means of convenience. Do they hide issues we need to be thinking deeply about? Or do they unnecessarily exacerbate our fears? I mean, generally this stuff is all taking place subconsciously. There's no attempt to expose these stories. That would be like doing psychoanalysis on yourself, for God's sake!" Rejeski stops short, as if momentarily surprised by the sharpness of his own analogy. But then he chuckles. "And there's definitely no one else up there talking about comic books. I'm almost embarrassed sometimes to be bringing that stuff to the table. I always imagine that there are people who are saying 'Oh man, I'm not going there. That's off the wall.'"

In the end, though, Rejeski seems to enjoy the idea of antagonizing people, just a little bit. "I've reached the point in my life," he reflects, "where I'm not particularly concerned whether I please the scientists or the policy folks. I think somebody's got to talk about this stuff, because it has huge implications that go right into the regulatory system." Think about that evolution, not revolution, storyline, he says earnestly. "That's one you really want to pull the veil back on. Scientists know that the wrong story will have direct links into regulation that they want to suppress."

If synthetic biology is seen as truly novel, Rejeski points out—if the narrative we tell about it resembles a science fiction plot instead of harking back to an old technology—then it will trigger the Toxic Substances Control Act and various Food and Drug Administration regulations. "These aren't," he concludes, "just superficial stories. So if I were 30 years younger and my career was at stake, I might be more sensitive; but now? I have no problem pissing people off if I think there's something that has to be said." When he says this, it's easy to imagine Rejeski as a character in his own compelling story. Not Pandora; not really. More like an 11-year-old boy holding up a homemade Geiger counter, using his own good sense to make invisible forces visible.

3

PAYING FOR PERENNIALISM

Sarah Whelchel and
Elizabeth Popp Berman

When Wes Jackson looks out across the wild prairie of Kansas, he sees far beyond the limits of his vision. "I'm surrounded by prairie here," he tells us, and the fields of waving grasses and deep roots have shaped how he thinks—about farming, about the relationship between our species and the planet, and about how to live.

Jackson speaks with a folksy authority, an assurance born out of decades of commitment to his cause. His words are calm and considered, but an underlying excitement buoys them along. Current agricultural practices are destroying soil, a precious resource that "we treat like dirt," and, unless we change them, this situation will only worsen with time. Jackson believes that perennial varieties of wheat, corn, and other crops that would not need to be planted every year could go a long way toward solving the problem. Perennials could be part of an agricultural system that tries to mimic the natural economy of an ecosystem such as the prairie, with its remarkably healthy soil. But first we'll need economically viable perennial varieties of major food crops, an ambitious goal. Jackson, who is 75, has been working to promote perennial agriculture for 35 years through an organization he founded, the Land Institute. He knows he may not live to see the fruits of his labor, but this doesn't trouble him. He insists we have an obligation to take on projects larger than ourselves. "A lifetime is a narrow interest," he says. His own interest is as wide as his beloved prairies.

Doomsday warnings about humanity's future follow a familiar narrative: The global population is growing. Greater demand is leading to the cultivation of marginal lands, which are less productive and more quickly exhausted. Overfarming causes erosion and the degradation of even high-quality soil, while nitrogen runoff from fertilizers expands dead zones in the rivers and oceans. And all this occurs against a backdrop of climate change. The status quo, it would seem, leads to catastrophe.

Jackson, like a prophet crying out in Kansas, provides an alternative vision: It will require real change, but these disasters can be averted, if we will take nature as our model of how to treat nature. The increased use of perennial crops, both those in existence and new ones yet to be developed, is a key component of this. The advantages of perennials over annuals (see the text box on the next page) lead to the tantalizing promise of a more sustainable agriculture system, increased long-term soil health, and greater food security. Creating these new varieties, particularly major cereals, is not a trivial task. But developments in the past decade—increasing numbers of interested scientists, faster and cheaper genetic sequencing technologies, helpful knowledge gleaned from research on perennial biofuel grasses, and increasing concern about environmental deterioration—have made perennial crops both more important and more realistic.

Why Perennials?

Many plants are naturally perennials, but grain crops by and large are "big-bang annuals," as Steve Jones puts it, that put a huge portion of their resources into producing seeds before "dying very efficiently." The potential advantages of perennial grain crops have been explored in depth elsewhere, but we recount them here briefly:

Reduced inputs. Seeds don't have to be planted every year. Perennials are more efficient than their annual cousins at absorbing nutrients, meaning that fertilizer inputs (and runoff) are reduced. Increased efficiency in using precipitation, combined with natural drought resistance, mean that less irrigation is needed. And natural pest resistance allows for the use of fewer pesticides.

Erosion control. Not planting every year means less tillage. Also, perennials put down deeper, bulkier root systems than annuals, helping to hold the soil in place.

Soil health. Existing perennial grass systems have much healthier soil than land under conventional or organic cultivation.

Carbon sequestration. Thanks to their much larger root systems, perennials naturally sequester more carbon. The potential for better yields Perennials have a longer growing season than annuals and don't have to devote resources to building a new root system each year. So although some worry that perennials will never yield as much as annual crops, other scientists believe they could one day actually produce superior yields.

But research on perennial crops suffers from a critical problem: lack of funding. Washington State University professor Steve Jones, who a decade ago helped identify a single gene that tells annual wheat to die, has shut down his laboratory genetics work almost completely. Other researchers confirm the challenges of finding support for science that strays from the well-worn path of agricultural research and production supported by the U.S. Department of Agriculture (USDA).

Perennialism research is risky, and its payoffs are weighted toward the long term. For the agriculture industry, it is not an obvious place to invest. Perennial crops would be disruptive to the current business model, since one of their key benefits is requiring fewer inputs—seed every three to five years, less fertilizer use, less pesticide use. Industry might respond to demand from farmers or consumers, which is how organic foods gained a foothold. But this can't happen until perennial varieties with decent commercial potential actually exist. Even the most optimistic projections predict that we are still 10 to 15 years out from this goal. Lack of incentives and a daunting time horizon mean industry is unlikely to be the catalyst for perennials.

Perennialization presents different problems when researchers seek federal support. The USDA has long been focused primarily on increasing yield, a metric where perennials face significant disadvantages. Annual crops have a head start of thousands of years of selective breeding. Also, many researchers assume there's a tradeoff between yield and perennialism: If plants put their resources into deep roots, those resources can't go into producing big seeds. Other scientists now challenge this assumption, but there's no question that starting from behind makes perennials less appealing to federal funders.

Beyond the yield question, alarmingly few resources at any federal agency go to supporting long-term, high-risk, high-reward research. The three-year grant cycle, with its emphasis on incremental improvements, just doesn't lend itself to this kind of science. Vision as far-reaching as Wes Jackson's may always be rare. But even a modest shift toward longer-term thinking could make a huge difference in our progress with perennialization, as well as with other projects that languish because of their timeframes.

We spoke with a number of scientists, policymakers, farmers, and activists about the growing interest in perennial crops, and the challenges of finding resources to support research. Three in particular serve as exemplars of what's happening in the field, in terms of both progress and problems:

- Wes Jackson, native Kansan and founder of the Land Institute, has been working to encourage the development of perennials longer than anyone else in the field.
- Steve Jones, a locally minded wheat breeder, has devoted himself to Washington State agriculture as he tries to answer some of the most elegant questions in biology.
- Ed Buckler, a programmer/geneticist who now regularly finds himself planting corn, wants to bring the incredible advances in DNA sequencing technology to bear on the problem of perennialism.

Wes Jackson

In 1940, U.S. wheat growers could expect to produce about 14 bushels per acre harvested. Decades later, that figure has tripled. The increase in corn yields is even more impressive: 10-fold since the 1930s. The success of plant breeders' efforts to increase yield is abundantly clear: Food production has kept pace with vigorous population growth. It has come at the cost, however, of extreme demands on the land itself.

Wes Jackson was born in Kansas in 1936, so he has observed many of these changes, good and bad, firsthand. His epiphany came in the late 1970s, while comparing present-day erosion with that of the Dust Bowl years. Despite measures undertaken to fight erosion over those intervening four decades, improvement was minimal. Perhaps, he realized, the problem wasn't something that farmers had actually tried to fix during those decades. Perhaps the problem was a much older, more fundamental one.

In Jackson's view, our most essential nonrenewable resource is not oil, but soil, and our present agricultural system is destroying it with industrial efficiency. When he talks about the "10,000-year-old problem of agriculture," Jackson's enthusiasm is captivating. As he describes the glaciers of the last ice age receding, and the planting of the first wheat crops near the Tigris river, it's easy to feel that agricultural history holds important secrets. When you grow annual monocultures, Jackson explains, as we have ever since those first wheat crops, you have to destroy the local ecosystem "to stay ahead of the weeds." And in fact, soil erosion has been around as long as farming; the Fertile Crescent today is a far cry from what it once was. For more than three decades now, Jackson has been working to raise awareness of this issue and to promote an alternative vision of how we should, or perhaps must, learn to farm if we want to feed the world without destroying it.

In 1976, Jackson founded the Land Institute in Salina, Kansas, to "develop an agricultural system with the ecological stability of the prairie and a grain yield comparable to that from annual crops." The prairie is held up as an ideal, at least in part, because it's where Jackson grew up and lives. "If you look out at the prairie," Jackson says, "whether it's tall-, mid-, or short-grass prairie, you don't have soil erosion beyond natural replacement levels. The only thing the prairie doesn't produce is nice big seeds."

Jackson, who holds a Ph.D. in genetics, advocates for a system of cultivation that evolved over millions of years: nature's. Instead of planting monocultures, he argues that we should plant multiple crops together in mutually beneficial combinations, or polycultures. And instead of replanting annual seeds every year, these polycultures should feature perennials with the potential to produce for three years or more. He calls this system perennial polyculture. If we're willing to mimic nature, we will, he believes, find inherently sustainable ways to farm, and the development of perennial grains is the scientific key to this task.

The Land Institute has been building support for this vision for 35 years, through advocacy and collaboration with scientists. In the past decade, though, frustrated with the slowness of progress, the organization has turned to doing plant breeding itself. Its staff scientists and fellows have since made significant strides toward perennialization, working with crops ranging from wheat to sunflowers to sorghum. Jackson's colleagues are now making baked goods from Kernza (the name is a play on Kansas kernels), a perennial intermediate wheatgrass they've developed. And Jackson playfully describes a field of annuals planted in polyculture that were harvested together as "instant granola." Designing machines to separate crops harvested together would be a straightforward engineering problem, Jackson says, so no need to worry if granola isn't your thing.

Today, the Land Institute's work is at a tipping point. With increased press attention over the past few years, public interest in the project of perennialism is beginning to build. During her tenure as Deputy Secretary of Agriculture, Kathleen Merrigan brought up the promise of perennials in several public forums, so they have had a champion at the USDA. But although the Land Institute's in-house plant breeders and the academic scientists across the country who are part of their network are making real progress in developing perennial food crops, their work is small in scale and poorly funded.

The Land Institute has doubled in size since its plant breeding efforts began, but raising money is an uphill battle. When asked how the organization is supported, Jackson deadpans: "Bake sales. Church socials." Though the organization has received small amounts of federal funding via earmarks and the USDA's Sustainable Agriculture Research and Education (SARE) program, the vast majority of its support comes from foundations and individuals. Jackson is hopeful that government will play a role in making his vision a reality. He has promoted a 50-year farm bill in Washington as an alternative to the existing five-year version, but his own current efforts focus on foundations, which could invest on a scale as large as his dream.

He's not the only one who thinks that foundations, rather than the government, are the most likely solution to the funding problem. Rick Welsh, a professor of sociology at Clarkson University who specializes in agriculture and food systems, agrees that "if you could convince someone at Rockefeller or Kellogg that this was really the future for the developing world or the United States, then maybe you could get a champion." Although they can sometimes be fickle in terms of their priorities, foundations do have the flexibility to think long-term if they choose. Cornell plant geneticist Ed Buckler points to the Howard Hughes Medical Institute, which supports exceptional biomedical scientists, not for a specific project but to pursue whatever research they think is most promising, however risky, for a renewable five-year term. And he observes that many leaders of recently established foundations "got wealthy by being in this high-risk venture. They think, 'Well, odds of this working? Maybe it's 10%. But if it does work, the payoff is massive.'"

Like his agricultural vision, Jackson's goal for foundation support is ambitious. He imagines a handful of big foundations partnering together to make a transformative

30-year commitment of around $3 billion. While aware that this is a long shot (at one point he assures us dryly, "I'll try not to have too much senile rapture here"), he also has a ready counter for those who might call his ideals utopian: "If you're working on something you can finish in your lifetime, you're not thinking big enough."

Steve Jones

Like Wes Jackson, Steve Jones worries about soil erosion and sees perennialization as a key part of the solution. But as an active scientist and wheat breeder, his focus is more on the genetic than the societal level. Jones works in the Skagit Valley, a mild and fertile region 60 miles north of Seattle and less than 10 from the Pacific Ocean. Here, spring brings acres of tulips and autumn more than 90 different crops, from apples to zucchini, harvested on farms that are typically smaller than 150 acres. Today Jones directs Washington State University's Mount Vernon Research Center there, but he started his time with the university at Pullman, in the southeastern part of the state, a dramatically different landscape.

Eastern Washington receives so little water that, as Jones puts it, "There are two crops: wheat and dirt." It was in this dry, windy region, where farmers struggle with soil erosion across many hundreds of acres, that he first began working on perennial wheat. He arrived at Washington State in 1991, fresh out of graduate school at the University of California, Davis, and soon started hearing from growers that they needed perennial wheat to combat erosion. Jones was vaguely aware that a professor named Coit Suneson had worked on perennialism at Davis decades before, but had no other knowledge of the subject.

It was a fortunate accident that led him, in the mid-1990s, to start thinking seriously about perennialism.

Jones had been using natural crosses to transfer genes for disease resistance from wild plants to their agricultural relatives. A couple of months after the harvest, he noticed that whole regions of a field planted with these crosses were growing back. Inadvertently, he had on some level transferred the perennial nature of the wild varieties to their annual cousins.

Serendipitously, around the same time, Jones received a call from a local farmer, Jim Moore, who was serving on the Washington Wheat Commission, a growers' organization. Moore had been asking about perennial wheat for years—he wanted something that would hold his soil down, and also be productive—but Jones was the first person not to laugh at his request. Moore suggested that Jones apply for support from the new Fund for Rural America, a USDA program created by the 1996 Farm Bill. Jones, who had not previously sought such a large grant, asked for $700,000 to fund four years of work on the feasibility of perennial wheat. He soon received a call congratulating him on the excellence of his proposal. The fund would support his research, but could only provide $500,000, and suggested that he do three years of work rather than four. "It was funny, given that we were working on perennial wheat," Jones notes, "but that's what we did."

Developing a new annual variety of wheat can take 10 years, but perennials have an even longer timeframe, because it takes several years, not just one, to see how they fare. Moore, who helps test Jones' experimental varieties on his farm, sees this as perennial wheat's biggest challenge: "The time horizon is just too long for people to be willing to support it." At 73, he too thinks he may not live to see perennial wheat become widely available, but he believes his granddaughter will.

Jones's work contributes to that hope. "It's a beautiful biological question," Jones muses. "We've bred wheat to be a big bang annual, which means that it dies very efficiently. What if we can tell it not to die?" One of his Ph.D. students was able to identify the gene or genes that causes annual wheat to die—the gene his inadvertent perennials seemed to lack. But simply transferring that one gene won't produce commercially viable plants. "There's tremendous work to be done there, just careers full of work," Jones says, a little wistfully.

The Fund for Rural America disappeared with the next farm bill. After that, Jones received three years of support from Western SARE, but since then he has had little success in winning competitive grants to work on perennialism. For a while, Jones and colleague Tim Murray, a plant pathologist, had a very modest federal earmark for their work, but this too dried up. Lack of funding has led Jones to terminate his laboratory genetics efforts, and he struggles to support graduate students. With additional money, Jones says, he would relaunch his laboratory work, as well as drastically expanding the scale of the perennial breeding program. "Breeding is about massive numbers, and that just takes people. Can we look at 100 lines or can we look at 50,000 lines? We have the space, we have the equipment, but it takes people to go out and do it."

To cope with limited funding, Jones has developed a different strategy for maintaining research support. He is still pursuing competitive USDA grants, but increasingly he is shifting his focus. "Our strategy is to decentralize our funding and get it locally. One of my Ph.D. students is funded by the Swinomish tribe here. He does water-quality and salmon work. A local town has given us $150,000 to fund a Ph.D. student. It's not for perennial wheat, it's for composting biosolids. So, we're being creative in that way about getting our students funded." Growers in the area are committed to Jones and to the research center. They donated the land it was built on all the way back in 1943, and they raised funds to rebuild it in

2006. The local strategy is working, in terms of supporting a productive research operation. But it means that Jones' research on perennial wheat is moving at a crawl, not a gallop.

Ed Buckler

When a scientific project is too premature for industry interest, and foundations haven't stepped in, the public sector is left to stand in the gap. So far, it's fallen short in this role with perennialism, thanks to historical tendencies in public agricultural research to fund short-term research at the expense of lengthier projects, and to focus on year-to-year increases in yield. Because funding for research on perennialism is so difficult to get, Jones no longer sees federal support as terribly promising for perennial crops.

Ed Buckler, a plant geneticist based at Cornell University, has a slightly different perspective. Buckler works at Cornell, but he is also on staff with the Agricultural Research Service (ARS), the in-house research branch of the USDA, which gives him a helpful perspective on what a public servant can currently hope to accomplish. Buckler works on maize, not wheat, and was raised in Arlington, Virginia, a long drive from farm country. He's been programming since he was eight or nine, and his approach to the problem of perennialism is rooted in computational genomics. Buckler is aware of the constraints on public support for perennialism, but he's doing his best to work within them and has some suggestions for reducing them.

Buckler's scientific goal is to speed up the process of developing perennials. He speaks fondly of Jackson's Land Institute: "They have been championing perennials for a really long time. The basic concepts that they spoke about a decade, two decades ago, are exactly right. But they would lay out a time horizon on the order of 50 or 75 years. That was beyond the normal attention span of what society wants to do." Today though, with the aid of genetic sequencing technology, Buckler believes that some of the Land Institute's goals could be accomplished much more quickly.

Buckler wants to accelerate the crossbreeding process by using DNA sequencing tools to identify the most promising naturally occurring variations. The technologies used to sequence genomes have advanced by leaps and bounds over the past few years. Sequencing the first human genome in the 1990s cost $3 billion; today, the price to sequence a genome will soon approach $1000. A project that would have been "a moon shot" fairly recently is now, Buckler explains, "something that could be done with a couple of million dollars a year." This shortened timeline opens up more potential for doing the work with federal funding. "I think we're at a point now where in one five-year chunk we could figure out the genes that are important for perennialism, and in another five-year chunk we could start putting those together to put a perennial out in the field. Not one that a farmer would care about, but one where biologists would say, 'Yes, that's a perennial, that looks like corn.' *If* we were successful." From there, it might be five more years to a plant that farmers might actually want to grow.

Still, even that plan assumes that five-year grants are available. But it's only recently and in a handful of areas that the USDA has given grants longer than three years. Reflecting on perennialization, Buckler says, "I think it will be interesting to people. I personally am definitely interested in doing this. But I know of [only] one source [the National Science Foundation's plant genome program] that would fund five years on something like this. Out of all the federal research portfolios that are likely to support something like this, there's one program. And that's kind of the problem."

Even a five-year time horizon is challenging for plant breeders, whether they're working on perennialism or other traits. "A realistic number?" asks Buckler. "For a plant breeder, it probably should be a 10-year grant, and maybe at two intervals you provide rigorous peer-reviewed progress reports. But nobody has 10-year grants."

Buckler's own work bears out the point that plant breeders need more than three years. "Some of our most successful and highest-profile projects actually took about eight or nine years to do," he notes. He credits his status as an ARS employee with allowing him to undertake such projects. His position "provides a core level of basic support—enough to hire a field manager, and a lab manager, and a couple other people. Compared to a regular academic position, that's a real advantage." That core funding has been critical to his research. "We used our hard money support from the ARS to set up the experiment over several years, and then we wrote a grant saying, okay, in five years we're going to finish this thing off." Scientists employed solely by universities, of course, cannot use this strategy.

The Question of Yield

But even in a world of 10-year grants, proponents of perennials would still have to address the second of these twin problems: yield. Even some supporters see perennials as unlikely to ever yield as much as annual crops. Bill Beavis, a geneticist at Iowa State University with an industry background, would like to see perennials succeed. But with seed companies investing hundreds of millions of dollars in improving their product each year, it's hard for him to imagine perennial corn ever catching up. For Beavis, the most difficult task will not be the scientific one of developing the corn, but the economic one of creating incentives for industry to invest in improving perennials' dollar-per-acre return.

But Buckler has higher hopes. He thinks perennial yields could potentially surpass those of annuals. "Some of the best data now comes from the biofuel efforts. In central Illinois, if you grow perennial biofuel grasses side by side with corn, the biofuel grasses fix 61% more carbon per year than the corn does," largely because of their longer growing season. "If we can divert that carbon not just to stalks but to kernels and ears, which I think genetically is totally doable, then you could make an argument that we should be able to beat the yield of modern maize by 60%."

In the long run, Buckler thinks industry will find a way to profit from perennials. "Seed companies spend a lot of money actually making seed, about a third of their costs. And so they could perhaps have better profit margins if they were on a perennial system that rotated seed every five years." He adds, "If we do our job right as geneticists at making perennials, they'll be so attractive that a lot of people in industry will take them up. But there needs to be enough basic research to push it to the point where we can say, 'Now we're within a five-year time horizon to take this to commercialization.'"

Only time will tell whether perennials can approach or surpass the yield of annuals. But Steve Jones also offers a reminder: "If it doesn't yield as well as annual wheat, especially at first, we're not all going to starve."

Admittedly, it's hard to see increasing yield as a bad thing. Agricultural land is finite; the global population is large and growing. Feeding the world is going to require us to produce more food, and proponents of perennials quickly run into this reality. "There's almost an obsession with yield," observes Leland Glenna, associate professor of rural sociology at Penn State University. "Getting *x* bushels per acre can even matter more to agricultural scientists than the money farmers make." Perennials have many advantages, but as of yet yield is not one of them. Other benefits, like reduced inputs and erosion control, are less resonant. Rick Welsh, Glenna's collaborator, adds "It can be hard for plant breeders to find funding or private sector collaborators if their innovations don't increase yield, even if they have other benefits."

The problem here isn't that yield is unimportant, it's that we think about yield on too short a timeline. Jim Moore points out that much of his land in eastern Washington is so dry he can plant wheat only every other year. "The question is," he says, "what would you accept in order to be able to plant perennial. Forty bushels? Thirty?" For him, 30 bushels every year would be coming out ahead.

And that's with a shift to a two-year horizon. Over a period of decades, as fertile topsoil blows away or runs off, damage to farmland can be irreparable. The strategy that maximizes yield over the next year, or even decade, may look very different from the one that maximizes it over the next 50 years. The United States is blessed with an abundance of fertile land, but even here erosion and constant cultivation are steadily degrading soil health. In countries where a significant portion of cultivation already occurs on marginal land, the situation is even more acute. To use an agricultural analogy, focusing too exclusively on short-term yields may be akin to eating the seed corn.

The Long View

Scientists interested in perennialism cope with the current funding situation as best they can, patching together bits and pieces of support or doing a little work on perennialism on the side of their main projects, as Jones has done. Limited resources do encourage deep thinking about what research is most meaningful and how to carry it out most efficiently, leading to efforts such as Buckler's. And even in this challenging environment, researchers are making progress: Jackson's Kernza could be commercially viable within a decade. Perhaps, particularly as rapidly improving DNA sequencing technology speeds things up, they'll pull this off even without easy access to funds.

Right now, though, funding priorities mean scientists are turning their energies in other directions. Steve Jones can't support laboratory research on the genetics of perennialism. Ed Buckler would like to work more on it, but already has a major grant from the one source he knows might support such research and can't apply for another. And when researchers turn away from a project, continuity, which is critical to plant breeding research, is lost. Jones recalls the work done at UC Davis: "When Coit Suneson retired in '61, a little of his material was saved, but the rest of it was thrown away. And then you start from scratch."

Some of the problems hindering perennialism's progress can't be solved. Sometimes there are simply more quality proposals than available funds. And in times of economic uncertainty, scientific research is not immune to budget cuts. The argument for perennials, however, is clearly strong. And it's made stronger when one realizes that perennial research has significant implications for annuals as well. Buckler explains: "Even if we fail at figuring out the genetics of perennialism, we would at least learn how cold tolerance works, and how drought tolerance works, and flooding tolerance, which would all be great traits to get into annual corn."

Wes Jackson, who has carefully pondered our aversion to thinking long-term, jokes that he's considered selling an American doll—you wind it up, and it gets bored. (All proceeds would go to the Land Institute.) Currently, though, interest in perennial crops is building, and there are a number of ways policymakers could act decisively to capitalize on this. First, Congress could simply name perennial crops a high-priority area for competitive grants in the next farm bill. Second, even without congressional action, the ARS could prioritize them as a research area. Welsh observes, "The USDA, through the ARS, has the resources and expertise. It makes sense for them to put money into this sort of public good. That's where you traditionally have seen research and varieties produced that are not necessarily going to be driven by the private sector."

Jerry Glover, a Science and Technology Policy Fellow of the American Association for the Advancement of Science and veteran of the Land Institute, acknowledges, however, that there is a challenge here: "ARS scientists, I think, would greatly welcome the opportunity to work on perennial grain crop development given more funding. But of course we know what the budget situation is. When you have to make difficult choices the public is in general going to support the more charismatic or compelling research programs versus food. Hopefully we can communicate the linkages between farming, the environment, and our own national security needs."

More broadly, a government reinvestment in plant breeding would benefit not only perennials, but a host of other projects that are too long-term to be on industry's radar, or that are of public value but not commercially viable. For a decade, agricultural scientists in the public and private sectors have been decrying the decline of public plant breeding. USDA support for plant breeding declined slightly in real terms between 1985 and 2005; the budget of the National Institutes of Health (NIH), by contrast, tripled during the same time period. Glover emphasizes just how critical such an investment is: "The number-one thing the public could do to adapt to and help mitigate climate change, and to ensure a food-secure future for the planet, is to reinvest in publicly funded plant breeding programs."

And at the most general level, both science and society would benefit if funding agencies prioritized identifying and supporting high-risk research of long-run importance. Academics can be a surprisingly conservative group when it comes to making grant decisions. Funding panels often go for the sure project over the unfamiliar, riskier, but potentially higher-payoff proposal.

Federal agencies are, of course, aware of this and slowly making efforts to counter it. The National Science Foundation, for example, has recently worked to identify and encourage "potentially transformative research" by establishing its Emerging Frontiers in Research and Innovation program. NIH has the Pioneer Award program to support innovative high-risk research by exceptionally creative scientists. At the USDA, a few large, five-year, multi-institutional projects are being supported. Expanding this approach would benefit not only perennial crops research but also other risky but high-impact research areas that are currently being neglected.

The main problem the scientists interested in perennialism face is clear enough: lack of funding and particularly of long-term grants. For the time being, industry isn't going to pay for the development of perennial grains. And although foundations could be the solution, so far they haven't seriously invested. But both the possibility and the need are clear. Occasionally, we need to leave the confines of our workaday lives, step outside, and join Wes Jackson in peering toward the horizon. The federal science enterprise should draw on this sort of vision and provide support for some bold long-term projects such as the breeding of perennial crops. The United States is a large country, and it can afford to spread out a bit of risk across so many amber fields of grain—especially when the payoff is protecting the fertility of those fields for generations to come.

4

THE LITTLE REACTOR THAT COULD?

Ross Carper and Sonja Schmid

A week before Halloween 2009, John R. Deal, an entrepreneur who goes almost exclusively by "Grizz," took the stage at the Denver Art Museum to deliver the headline talk at an evening seminar titled "The Truth About Nuclear Energy." Though slightly less bearded, barrel-chested, and commanding than his nickname suggests, Deal's style was exceedingly casual. A long-sleeved, amply pocketed khaki shirt included a shoulder patch with his company's logo, eliciting an association somewhere between park ranger and scoutmaster—both of which match his cheerful, disarming demeanor. Before launching into the benefits of his company's miniature nuclear reactor, he began with a joke.

"It turns out that most of the... mishaps [in nuclear plants] actually involve humans. So we were thinking today, what do we do to create a power plant control system to minimize that kind of impact? We came up with the following. The power plant of the future will have three control devices: a computer, a dog, and a guy. The computer runs the power plant because, as I said, most power plant mishaps happen because of human interaction. The dog keeps people away from the computer. And the guy is just there to feed the dog."

After lingering on the title slide a moment longer—"New. Clear. Energy." in yellow letters—he advanced the screen and gave his opening line, a message he would revisit throughout his talk. "It's more of a battery metaphor."

As the co-founder and president of Hyperion Power Generation, Deal was referring to his company's starring product, which he believes will represent a radical revolution for nuclear power. He has also described the Hyperion Power Module (HPM), which is only a few feet wide and not much taller, as the iPhone of nuclear power: a compact, technologically elegant device that will be a worldwide sensation for its portability, ease of use, and applications. These first moments of a normal overview presentation contain two of Hyperion's prominent talking points: a piece of imagery and a problem solved. HPMs are batteries that eliminate nuclear energy's obstacles related to human error and expertise. For the latter point, his Denver talk and many others refer to the goal of taking Homer Simpson out of the equation.

When Sonja Schmid and I set out to capture the story of small modular reactors, it quickly became clear that this technological coming-of-age tale is really, at least for now, a story about stories—the imagery industry leaders use to both envision their designs and communicate them to policymakers and the public. Behind the technical fact sheets, and in the years that remain before designs become physical machinery, small reactors are a movement of metaphors.

On many topics, imagery doesn't carry substantive weight. It is added for flavor, to simplify, clarify, or restate content in more vivid terms. But in the house of small nuclear reactors, metaphors seem to be weight-bearing walls.

They also come in the context of a debate that couldn't have higher stakes. On one hand, our world must quickly scale up new sources of carbon-neutral energy. On the other, the nuclear accident in Fukushima, Japan, reminded us that our attempts to do so in the nuclear sector may result in unforeseen complications that can spiral into disasters. In today's proposals for a new nuclear approach, presentation matters. But how much does corporate imagery reveal about the technology itself and its implications, and how accurate are the pictures the industry paints?

Is Small Beautiful?

Overall, the emerging vision of small modular reactors is a major downshift from the custom-built giants of yesteryear to new railcar-ready, factory-manufactured, standardized machines with an electricity output in the range of 25 to 200 megawatts (MW), rather than the 1,000 or more MW that is typical in today's commercial reactors. A growing faction of promoters believes that these small reactors can provide solid answers to the myriad risks nuclear energy continues to face: safety, weapons proliferation, waste management, and initial capital cost. Each small reactor design offers a unique narrative of how it will remove or reduce these risks. Recurring themes include built-in capsule-like containment, passive cooling features, pledges for more effective disposal or recycling of waste, and a kind of inverse "economies of scale": advantages offered by small capital investment, standardization, and mass production.

Because none of these small designs has yet been licensed by the Nuclear Regulatory Commission (NRC), and all of them are still several years from market deployment in even the most optimistic scenarios, they make a convenient canvas on which to paint metaphors.

In the case of radically advanced reactor designs and deployment strategies, both corporations and journalists readily put vivid colors to use. Others are cast in more muted, evolutionary tones as miniature versions of the world's tried-and-true light-water reactors, with substantially improved safety features. Leading revolutionary approaches in fuel, moderation, and cooling include reactors by Hyperion, Toshiba, and GE Hitachi, whereas efforts in favor of a more incremental design change include NuScale, Westinghouse, and Babcock & Wilcox.

All leading small reactors create a modular option, which allows them to be pieced together like LEGO blocks to build up a customized power supply. Customers could potentially receive their prepackaged mini-reactors anywhere in the world, as long as the site is accessible by boat, truck, or rail.

Judging by a rising emphasis on small modular reactors within President Obama's past two budget requests, not to mention former Energy Secretary Steven Chu's outspoken affection for the technology, small reactors are increasingly being considered a highly exportable clean energy innovation and therefore prime candidates to implement the administration's "win the future" message.

Returning to Hyperion, the way they present their technology shows that subtlety is not a priority. In some sense, there is a space for this; the small reactor market is already revolutionary in that it allows room for entrepreneurs to join the nuclear energy ranks alongside giant, buttoned-up corporations. And some entrepreneurs have a habit of making big, bold claims—early and often.

For example, a February 2011 *Time* magazine article titled "Nuclear Batteries" prominently features the "tanned and enthusiastic" Grizz Deal. Curiously, the author of the piece uses the phrase "nuclear battery" throughout, not as a metaphor but as the default label for Hyperion's small reactor. Along the way, Deal outlines his goals for the HPM, a commercialized design that is based on work performed at Los Alamos National Laboratory. By the end of the article, he is quoted offering to "take care" of much of the world's nuclear fuel, precluding the need for new nations to pursue enrichment or reprocessing programs, because these countries will presumably rely entirely on leasing Hyperion's product.

The *Time* article is not an outlier. In dozens of trade and popular press articles, interviews, and blog posts, the character of Grizz and his imagery shine through. In November 2008, he was quoted in the *Guardian* on Hyperion's safety and nonproliferation features: "You could never have a Chernobyl-type event; there are no moving parts," said Deal. "You would need nation-state resources in order to enrich our uranium. Temperature-wise it's too hot to handle. It would be like stealing a barbecue with your bare hands."

Seeking out the origins of the venture helped us fill in some of the history behind the enthusiasm. It began with an initial shared motivation, which was recounted to us in an interview with Deborah Deal-Blackwell, Deal's sister and cofounder of Hyperion. "My brother and I—neither of us have kids," she said. "About five years ago, we started asking, what can we do to leave a legacy in the world? After some searching, we found that clean water was the answer."

Deal-Blackwell explained the leap from clean water to nuclear reactors. She and Deal had quickly found that providing clean water on large scales, such as through desalination, can be quite energy-intensive. So they began to explore options. After briefly looking into renewable energy sources, they decided on a nuclear solution to pursue their clean water mission. Deal had worked at Los Alamos as an entrepreneur in residence, and he knew of an advanced reactor design by the lab's Otis Peterson that he thought would be perfect to commercialize. The HPM concept was born.

Peterson's design was technically intriguing to say the least. It would use uranium hydride, a novel nuclear fuel with unique self-regulating features that control the core's temperature. But in 2009, foreseeing licensing delays with such a revolutionary approach, Hyperion decided on an entirely different design Los Alamos had produced: a uranium nitride–fueled fast reactor cooled by molten lead-bismuth. In other words, instead of forcing the NRC to create a new classification, Hyperion intends, for now, to fit its reactor within the somewhat more familiar, but still far from commercial, Generation IV category. Interestingly, the only previous application of a lead-bismuth cooled reactor was in the Alfa-class Soviet submarines developed in the 1960s.

The HPM is also revolutionary in its size and its approach to spent fuel. The smallest of the leading design proposals, each unit would produce 25 MW of electricity, enough to power 20,000 U.S. homes—or considerably more homes in any other nation. Also unique is the approach of providing a factory-sealed unit that would be removed completely for refueling and waste removal every five to ten years, alleviating proliferation concerns related to sensitive material accumulated in spent fuel. This is a clear innovation that, if successful, would be a positive step forward from traditional practice. As a result, the approach offers an advantage over other small reactor designs, which do not seem to contain substantively new solutions for dealing with the on-site accumulation of spent fuel.

However, returning to the notion of human expertise reveals a clear weakness. Deal-Blackwell also told a version of the "feed the dog" joke during our interview, a repetition that implies that, in Hyperion's view, human expertise is best handled by sealing it inside an automated technology. Although concerns about human error are legitimate, neither the public nor government regulators are ready to accept that scenario. Emerging technologies such as Hyperion's call for a new and robust regulatory plan to determine what kind of human expertise is necessary for their safe operation, as well as how relevant knowledge can be created and maintained, transferred when appropriate (such as during export), and secured from illicit applications.

For three years, the "battery" metaphor has been the centerpiece of Hyperion's identity. Although some of this language seems to have been scrubbed from the company's Web site, former statements are easy to find on other sites devoted to the leading edge of nuclear technology. One example, from an early Hyperion Web page, began with the text "Hyperion is different. Think Big Battery..." and ended with, "Think battery, with the benefits of nuclear power. Think Hyperion." With this direct exhortation to nontechnical audiences on exactly how they should think about a small reactor, Hyperion is unmatched in its brazen communications. And as the *Time* article shows, the image has stuck.

The question is whether it fits. In one way, it does. The HPM is envisioned as a self-contained sealed unit, delivered and used until its fuel has depleted, then carefully returned to a proper facility. But the comparison doesn't hold much further than procedural similarities. A battery is a static device that converts stored chemical energy to electrical energy. It arguably does not belong in the same conversation as harnessing a nuclear chain reaction, the results of which include highly radioactive materials. Images on Hyperion's Web site of buried, unattended nuclear reactors would make sense if they were merely batteries, but they are not. For this reason, more than one of the nuclear energy experts we interviewed used the term "fantasy" in reference to such scenarios that deploy "walk-away-safe" nuclear reactors.

In the middle of Deal's talk in Denver, he began flipping through some artist-drawn images. The most striking of all shows a small nuclear reactor, buried and unattended at what looked to be less than 15 feet below the surface. Two simple tubes snake upward from the reactor, drawing the eye to a pair of gray above-ground tanks, with the words "Potable Water" stamped on the side. The setting? An impoverished African village complete with about a dozen mud-constructed, thatch-roofed huts. A handful of people were drawn into the image, all of them walking to or from the clean water source, which is apparently powered by a $50 million HPM.

Although the humanitarian goals that launched Hyperion are admirable, this quaint portrait of a Third-World problem goes beyond vivid jokes, iPods, batteries, and barbecues to reveal a full savior narrative that casts Hyperion's small reactor as a solution to some of humanity's direst needs. And the message is reinforced again and again. A recent news article in South Carolina's *Aiken Standard* led with the following sentence: "Nuclear power is the only thing that can save the human race, Hyperion Power Generation CEO John 'Grizz' Deal told a crowd of more than 150 in Augusta on Wednesday."

A utopian narrative is not without precedent in the history of nuclear power. In fact, it harkens back to the early 1950s, when the American public first heard rumors that "atoms for peace" would soon yield "electricity too cheap to meter." Early in our search for the story of small reactors, we began to notice something familiar: The shift to small modular reactors has the nuclear industry playing out the plot of *The Little Engine That Could,* a slice of mid-20th-century Americana that became a hallmark of children's self-esteem building. Where the large have failed to try, or tried and failed, the Little Reactor will come along and prevail, pulling the heavy load of toys and goodies over the mountain. Or at least the Little Reactor thinks he can.

An Emphasis on Evolution

The Little Reactor character appears in many forms, most of which are far less colorful than Hyperion's version. We spoke to Bruce Landrey, chief marketing officer at NuScale Power, a small-reactor startup based in Corvallis, Oregon. Landrey has spent his career communicating information about nuclear reactors for various companies. The story of his experiences, at its end, harmonizes well with his current employer's approach.

When Landrey graduated from the University of Oregon in the mid-1970s, he didn't have a job, and he wasn't necessarily looking to go into the energy sector. But soon his father was paired on the golf course with a stranger from an electric company that happened to be seeking new communications talent for the rollout of a new nuclear power plant. Eighteen holes later, Landrey's father had positioned him, without his knowledge, as a prime candidate for the job. He applied, and was hired.

"I was thrown into the deep end," he said, remembering how little he knew about nuclear power. He also encountered an odd phenomenon related to public perception in his region. "We had a lot of protesters and demonstrations at the plant, people chaining themselves to the fence and so on," he remembers. "But it was ironic, because the protesters were the same people I was drinking beer with the previous year at the university. But here I was, on the other side of the issue."

Landrey decided that if he would be earning his living speaking in favor of nuclear power, he would use his first six months on the job to learn everything he possibly could about the technology and its implications. He did so, becoming immersed in the technical side of nuclear reactors enough to make him confident discussing them from an environmental and safety perspective.

"But what I was never comfortable with was the tremendous business risk a large nuclear power plant poses to an electric company, its customers, and its shareholders," he said. And over the next several years, he had a front-row seat to the downsides of this risk. "The company I worked for tried to build two additional nuclear plants, which became caught up in licensing delays. Then, after the Three Mile Island accident, they were finally just abandoned."

Three decades later, Landrey still finds himself speaking up for nuclear energy, but now for NuScale. He is as risk-averse as ever when it comes to the financial challenges presented by nuclear power. So is NuScale, and this perspective guides both its technical approach and its communications. As the company sees it, their strategy builds on proven market-ready technology, familiar to regulators and the community of existing experts. Compared to revolutionaries such as Hyperion, the essence of NuScale's metaphor is much less splashy: Our small reactor is really an improved version of the reactor down the road. It is a light-water design, which means it uses normal water as its coolant, and it shares this feature, along with standard fuel rods, with the majority of active nuclear power reactors in the world.

Landrey explained some differences between NuScale and its larger predecessors, while also evoking a metaphor: a Thermos. Rather than a large concrete containment building, each reactor module comes inside its own steel vessel, which performs the containment's safety purposes while also forming a Thermos-like vacuum between the vessel and the reactor module. This enables the reactor's passive cooling feature, which uses natural circulation by a convection process, eliminating the need for a normal light-water reactor's mechanical equipment or backup power generation to cool the reactor. Of course, backup power generation was the key failure that set off the Fukushima disaster and is the Achilles heel of all existing nuclear power plants.

When we asked about Hyperion and other small reactor designs, Landrey was quick to draw a line in the sand between NuScale and a less traditional approach. "You have to be very careful with small modular reactors," he said, "to distinguish what goes in the near-term commercialization category and what continues to remain a concept in a laboratory someplace. There is a big gulf—it's really apples and oranges."

He also mentioned key differences on the topic of human expertise. Rather than automation, Landrey spoke of the importance of education and training in any context that will use NuScale reactors. The company's plans call for an expert staff to operate the facility. For example, the top image on the company's "Our Technology" Web page is an overhead view not of a reactor itself, but of the control room and user interfaces for plant operators.

For Landrey, the evolution-versus-revolution question is a central issue to explore when looking into small reactors: Which designs, or aspects of the design, grow out of widely used commercial power reactors, and which represent completely new attempts? The unstated perspective is that the evolutionaries represent realistic near-term solutions, whereas the revolutionaries are still far more futuristic than their promoters will admit.

Dusting Off a Design

Also quick to emphasize this gulf is Babcock & Wilcox, one of the world's preeminent suppliers of nuclear reactors. B&W is now partnering with engineering and construction giant Bechtel to develop and produce the "mPower," a compact new light-water reactor similar in many ways to the NuScale design. In the summer of 2010, Christofer Mowry, president of B&W, told the *Wall Street Journal*, "Bechtel doesn't get involved in science projects. This [agreement] is a confidence builder that the promise of this small reactor is going to materialize." Of course, as with Landrey's comment, such a quote cleverly forces the question into the reader's mouth: Which of today's small reactors should be dismissed as mere "science projects"?

Although the mPower is certainly an advanced project, its first draft has been around for quite a while; our interviewees spoke of their small-reactor effort beginning by "dusting off a technology from the early eighties." Compared to a conventional pressurized water reactor, the mPower reactor has the distinction of integrating the entire primary system (the reactor vessel, the steam generator, and the pressurizer) in one containment structure, which, according to one of the B&W engineers, "gives us a lot of inherent safety features that the large reactors don't have."

The tendency to look backward before moving forward arose not only from B&W's vast experience with light-water designs. First, it was a conscious response to its perception of the market. Many potential mPower customers are utilities that run today's fossil fuel plants (not exactly the most venturesome bunch), who will perhaps one day need to turn their turbines using a carbon-neutral technology. Hypothetically, a significant number of these utilities that would be priced out of a large reactor would, in fact, be interested in a more manageably sized, and priced, option. This thinking was the result of an executive saying flatly "show me a customer" when the company's technical leaders approached him with their idea about a small-sized, budget reactor. But a related and perhaps greater motivation for B&W's design conservatism is the current regulatory gatekeeper.

"The Nuclear Regulatory Commission... is a light water reactor regulatory agency," one of our B&W interviewees said. "It takes a very long time to come up with a regulatory framework to be able to license another type of technology, and we wanted to get the technology to market as quickly as we could."

Another interjected, "The idea was to come up with a design that capitalizes on the tremendous knowledge base that surrounds light water reactors, and then make some evolutionary changes. But when you get into revolutionary changes, the market isn't looking for that right now."

The design includes a plan to bury the mPower underground. Although this feature is widely shared across the small-reactor industry, B&W offered an interesting reason when we asked why. They first referred to aesthetics; their initial rationale had been to avoid the stigma associated with the physical appearance of a nuclear power plant. The typical cooling towers and containment structures have acquired almost emblematic status among opponents of nuclear energy. Only after having volunteered these reasons did they add that the underground placement also earned them safety advantages with regard to earthquakes and missile impact.

Like Mowry's reference to "science projects," B&W's presentation is subtle but quick to make use of the public's associations. Rather than taking a direct approach to force positive associations through imagery, B&W and others find the negative associations we already hold, and offer just the opposite. As they do so, the message comes back to their historical credentials, familiar technology, and the inclusion of credible players such as Bechtel. And the continuity of mPower's design sends its loudest message to the regulatory community: This is a well-known, mastered technology, but upgraded to add significant improvements.

The Appeal to History

Our foray into the light-water approaches coalesced in one question: Does inertia trump innovation in the U.S. nuclear industry? It would seem so, at least judging by NuScale's and B&W's carefully chosen paths. To some extent, even Hyperion's shift in reactor fuel for its initial small reactor sends a similar signal. A familiar picture emerges, where the very entities that serve as the guarantor of safety also represent an obstacle to new, potentially better ideas. Perhaps unintentionally, they provide incentives for companies to continue down the well-trodden path, in exchange for faster licensing approval and shorter time to market.

In terms of accounting for human expertise, evolutionary approaches do have a marked advantage. They do not seek a technical fix that eliminates the operator's crucial role and ignores organizational and educational structures. On the downside, however, slow incremental innovations tend to neglect nuclear energy's historical problems. The known hurdles with traditional light-water reactors, including low efficiency and unresolved waste management concerns, will arguably continue to live on for another generation, and if their industrial promoters get their way, these problems will be mass-produced and widely exported.

Other potentially valuable lessons from history are also ignored; for example, why there is so little commercial experience with small nuclear reactors. In the past, small reactors have been used in research settings, for naval propulsion, and, rarely, to power research or industrial facilities at remote locations. But until recently, most small reactors for research and on submarines and icebreakers operated on highly enriched uranium, material that in sufficient quantities could be used to produce a nuclear weapon. When converted to fuel with lower enrichment, these reactors require more frequent refueling. Furthermore, the United States abandoned small reactors altogether in the 1970s to take advantage of the anticipated economies of scale to be achieved with larger power reactors. As the story has gone, in many cases the word "economy" hasn't proven to apply.

In the 1970s and 1980s, the U.S. nuclear industry was embroiled in a debate over the safety of scaling up. Would substantially increasing the size of nuclear reactors allow extrapolation from existing safety protocols, or would it in fact produce qualitatively new problems? Similar questions should be asked in today's opposite scenario. It is far from self-evident that a compressed scale automatically produces smaller risks or that the data gathered from similarly fueled and cooled large reactors transfers down.

And if the evolutionary approach does lower the risk of a given small modular reactor, who can say whether reduced risks in individual power plants are outweighed by an overall global risk of dispersing a much greater number of nuclear reactors across the planet? The Fukushima disaster has inconveniently shown a problem inherent to installing multiple reactors at one plant. After a scenario of unique failures within several reactors at once, is the prospect of a dozen or more interrelated small modular reactors on one site still as attractive?

An overarching question is whether any of these risks are significantly curbed by an approach that offers familiarity, or whether this would encourage complacency. Pyotr Neporozhni, who served as the Soviet minister of energy and electrification for three decades, is reported to have dismissed concerns about nuclear safety with the quip: "A nuclear reactor is just another boiler." Neporozhni retired in 1985, one year before Chernobyl.

Although it is true that the end task is to boil water, it would be a mistake to ignore the intricate, wholly new ways in which small modular reactors will attempt to go about that task, even if widely known materials are used. A small design is not "just another light water-reactor."

Even if, as one B&W representative said, the NRC has traditionally been a "light-water-reactor agency," its leadership does not seem to be glossing over the novel questions small modular reactors are raising. During a summer 2010 keynote address at a conference devoted to small reactors, William Ostendorff, a current member of the NRC, indicated that the question is open regarding how much history counts toward confidence about new small reactors.

"There are substantial differences between the proposed concepts for SMRs [small modular reactors] and the large, light-water reactors that the NRC's regulations were based upon," he said. "How will prototype reactors be licensed? How will risk insights be used? How do SMRs fit into the Price-Anderson nuclear liability framework? Questions like these are not easy ones to answer."

Mixing Metaphors

During her dissertation research on the Soviet nuclear industry, Schmid spent a year in Moscow, mostly penned inside musty archival reading rooms. But with a single tape recorder and without a quiet office at her disposal, she also set out to preserve a primary resource that was, and is, dying out. Former dons of the Soviet-era nuclear power program spoke with her on trains and buses, in homes and coffee shops, and over sometimes-obligatory shots of vodka. One of these interviews yielded an image that stuck with her, a counterweight to the simplifying metaphors we had encountered.

Like her other interviewees, "Yuri" had been eager to speak to Schmid, but visibly relieved when she offered not to use his real name. For an elderly Russian nuclear engineer whose Cold War career had comprised stints in both military and civilian reactors, secrecy fell somewhere between a reflex and a superstition.

After two terse hours with her microphone on a desk between them, they shared a cigarette break. They stood in a stairwell, holding cigarettes over the public ashtray, a large metal trash bin painted, rather sternly, the same gray as the walls. Then, in two sentences separated by a narrow downward stream of smoke, Yuri abandoned his technical talking points.

"The reactors are like children; each one is different," he said, as if suddenly remembering something he had forgotten, the central point. "You come to know their peculiarities by spending time with them; you begin to feel how each reactor breathes."

The large traditional reactors he had operated during his career were supposedly identical in design, but as he said, their personalities were quite distinct, as if there was something immeasurably complex happening beyond the components of these machines, something relational.

Historically, nuclear energy has been entangled in one of the most polarizing debates in this country. Promoters and adversaries of nuclear power alike have accused the other side of oversimplification and exaggeration. For today's industry, reassuring a wary public and nervous government regulators that small reactors are completely safe might not be the most promising strategy. People may not remember much history, but they usually do remember who let them down before. It would make more sense to admit that nuclear power is an inherently risky technology, with enormous benefits that might justify taking these risks. So instead of framing small reactors as qualitatively different and "passively safe," why not address the risks involved head-on? This would require that the industry not only invite the public to ask questions, but also that they respond, even—or perhaps especially—when these questions cross pre-established boundaries. Relevant historical experience with small compact reactors in military submarines, for example, should not be off limits, just because information about them has traditionally been classified.

The examples we discussed show that metaphors always simplify the complex technical calculations underlying nuclear technologies. Vivid illustrations often obscure as much as they clarify. Small reactors are not yet a reality, and the images chosen to represent them are often more advertisement than explanation. What information do we need to navigate among the images we are presented with?

Clearly, some comparisons are based more on wishful thinking than on experience. A retrievable underground battery and a relationship with a child, for example, invoke quite different degrees of complexity. Carefully scrutinized, the selection of metaphors often reveals the values that go into the design of these new reactors: why one approach is safer than another, which level of risk is acceptable, and whom we should trust.

Ultimately, the images offered by our interviewees are based on projections. Although it may make intuitive sense that smaller plants will be easier to manage, nuclear power involves non-nuclear, and even nontechnical, complexities that will not disappear with smaller size, increased automation, or factory-assembled nuclear components. For instance, nuclear reactors, and by extension nuclear power plants, need reliable organizations to train experts, provide everyday operation and maintenance, address problems competently when they arise, and interact effectively with the public in case of an emergency. This is not a trivial list even for high-tech nations like the United States, and it presents an even larger challenge for prospective importers of small modular reactors, particularly the developing countries with no domestic nuclear infrastructure that are clearly a major target of Hyperion's efforts.

The same goes for the projected cost of small reactors. If there are any numbers publicized at this point at all, they tend to increase monthly, not least because of the recent events in Japan. The nuclear industry may need to rethink nuclear safety issues, revisiting problems it had considered long resolved. Small modular reactors do not offer easy solutions to multiple point failure. In fact, the modular arrangement of multiple cores at one site might increase this particular risk. These questions remain, regardless of whether a new reactor follows an evolutionary or a revolutionary track.

Whether the "nuclear battery" or the "just another light-water reactor" message appeals to us, we would be well advised to keep in mind the connotations of familiarity and controllability they offer in the face of unpredictable novelty. That should make us suspicious. Is what we are being sold as advantageous in fact the biggest vulnerability of these small designs? Easy transportability may look less like an asset when considered from the standpoint of proliferation. Multiple small cores might not necessarily turn out to be safer than one large one. We may remember that taking apart a machine, looking inside, and trying to figure out what is wrong ourselves can be more appealing than a machine that, like an iPod, needs to be shipped back to the factory for repair. Distributed generation sounds like a good idea when we talk about solar roof panels, but may not be as attractive when it requires highly trained expertise and accumulates radioactive waste.

We don't know all the answers yet, but we should avoid being drawn in too quickly by these metaphors, even those that are more muted than Hyperion's. Yuri's realization that reactors are like children, an image based on profound experience and devoid of any marketing bias, presents a different and competing picture. Rather than simplifying, Yuri's image goes in the opposite direction. Thinking of small reactors as more like children offers a lesson in humility in the face of complexities, both technical and nontechnical. Reactors, like children, may come with their own complicated personality; they can be quite unpredictable, but they also hold the promise of a better future.

Today's small-reactor narrative isn't a children's story but an immensely complex novel, rife with layers of context, relationships, and flawed characters. But even children's literature can temper itself against its own oversimplifications, as we are urging the nuclear industry to do. In 1974, Shel Silverstein published his reaction to *The Little Engine That Could,* flipping the empowerment narrative to a cautionary tale. The last stanza of his poem "The Little Blue Engine" warns against allowing confidence and optimism to become hubris:

He was almost there, when – CRASH! SMASH! BASH!
He slid down and mashed into engine hash
On the rocks below... which goes to show
If the track is tough and the hill is rough,
THINKING you can just ain't enough!

For the small-reactor movement to truly come of age, the metaphors we use to describe it must also mature. Convenient images, save-the-day narratives, and a we-think-we-can reliance on a purely technical fix must be balanced by a broader examination of a full range of metaphors, the complexities they capture or ignore, and the social, political, and organizational contexts in which these machines will ultimately be used.

Postscript, February 2015

Hyperion changed its name to "Gen4 Energy" in 2012; under new leadership, the company decided not to pursue a Department of Energy program designed to support SMR licensing. Gen4's new CEO, Bob Prince, argued that rather than focusing on the U.S. SMR market, Gen4 Energy's focus would "remain on regions and applications most in need of next generation technology." Commentators in the American Nuclear Society, however, suspected that Gen4's decision had more to do with the fact that a non-light water design (such as the Gen4 Module) would take much longer to get licensed by the NRC.

In 2014, Babcock & Wilcox "restructured" the mPower SMR program, that is, they slowed down the current pace of development as a result of not being able to secure enough investor interest. Provided that B&W's partners, including DOE, Congress, and the Tennessee Valley Authority, continue funding the SMR program, they plan "to move this technology toward licensing and deployment in the mid-2020 timeframe" but also warn that any such prediction involves significant risks and uncertainties.

As of February 2015, NuScale Power was testing a helical coil steam generator (HCSG) at a contractor test facility in Piacenza, Italy. The data produced by these tests are intended to validate NuScale's computer codes, and to optimize steam turbine performance. NuScale plans to submit a Design Certification Application to the NRC in 2016, and a Combined Operating License Application for the Utah Associated Municipal Power Systems Carbon Free Power Project in 2017, for commercial operation to start in late 2023.

Further Reading

Recent publications on social and symbolic dimensions of SMRs include:

Ramana, M. V., and Zia Mian. 2014. "One size doesn't fit all: Social priorities and technical conflicts for small modular reactors." *Energy Research & Social Science* 2: 115–124.

Sovacool, Benjamin K., and M. V. Ramana. 2015. "Back to the Future: Small Modular Reactors, Nuclear Fantasies, and Symbolic Coverage." *Science, Technology, & Human Values* 40 (1): 1-30.

5

LIVING AND BREATHING PLANTS

Angela Records with Roberta Chevrette

I peered closer at the brown lesions that appeared between the veins. An official diagnosis would take a few days, but it was clear that death was imminent.

Pleased with the results, I smiled.

On the stainless steel counter, lit by natural light from the row of windows that flanked one wall of the laboratory, were about a half-dozen clear, four-inch round plastic vessels, each containing a similar specimen of *Phaseolus vulgaris*—green beans. Though we are most familiar with the fruits of this plant, the leaf, too, is edible, and the dry stalks of the plant can be used as animal fodder. As a Ph.D. candidate at Texas A&M University, I was studying the disease known as bacterial brown spot, examining the processes by which it attacks the leaves, pods, and stems of infected plants.

Although I have never been overly fond of green beans, my satisfaction did not come directly from the fact that my specimens were dying. I found monitoring the small changes that occurred as the disease attacked the plants to be fascinating. More importantly, I knew that the results of my experiments on how the bacterial brown spot pathogen infects plants could help farmers and gardeners stave off the disease. I recorded my observations in my lab notebook and put the specimens back in the incubator.

Eleven years later, I have traded long days in the laboratory for equally long days working in and around the nation's capital. But my obsession with plant diseases remains intact.

I live and breathe plants. Technically, we all do. Without plants, we'd have no oxygen to fill our lungs, no clothing to fill our closets, no medicine to fill our bathroom cabinets, no food to fill our pantries. But how many of us roll out of bed in the morning (from under sheets and comforters made of plant fibers) and say, "Thank you, plants"? How many of us pull on a favorite T-shirt, sit down to eat a bowl of cereal, and reflect on the science that went into protecting the cotton and wheat crops from pests and pathogens?

Ancient cultures and religions often centered on agriculture. Festivals coincided with planting and harvest cycles, a tradition still carried on by many indigenous peoples around the world. Even many of the scientific names of plants still carry the mark of the ancient gods and goddesses with whom they were once associated. *Dianthus,* the scientific name given to the carnation, translates as "the flower of god," while *Theobroma,* the genus containing the cocoa tree, translates to "food of the gods." Zeus, Aphrodite, Venus, Artemis—they, too, show up in the scientific names of a variety of plants. *Agave,* the mythological namesake of the genus of spiny plants that includes those from which tequila is made, was a fierce goddess who tore her own son to pieces, believing he was a lion.

A continued tradition of agriculture worship would certainly make my current job easier; I'm a consultant for

Eversole Associates, a small agriculture, science, and technology advocacy firm that represents science groups in Washington, DC. If the Pledge of Allegiance were supplemented with, say, a prayer in homage to Demeter, the Greek goddess of the harvest, or to Chicomecōātl, the Aztec goddess of agriculture, plant science advocacy might not seem like such an uphill battle. But some days, I feel like members of Congress are Agave and the programs I fight for are the mythological son—unknown and about to be torn into tiny pieces.

***

September 7, 2011, was one of those days. Time was running out for Congress to pass the budget for the 2012 fiscal year, and the Hill had become the site of fierce and ongoing debates. I had been assigned the task of escorting a group of scientists from the National Plant Diagnostic Network (NPDN) to the Capitol, where we would meet with representatives from their home states to argue for the program's funding to be restored. The NPDN, a system of U.S. diagnostic laboratories established in 2002 to mitigate the impacts of plant pests and pathogens, had achieved a number of important successes in its nine years of existence. So NPDN scientists had been taken by surprise when, three months earlier, the version of the budget passed by the House zeroed out the successful program's already minimal funding. My boss, Kellye Eversole, the head of Eversole Associates and a former professional staffer for the Senate Committee on Agriculture, Nutrition, and Forestry, had received a call; could we assist the scientists in representing their interests to Congress? We rallied our forces and scheduled a number of meetings on the Hill for the four professors who flew in from various universities around the nation. But it was down to the wire—the Senate would be releasing its agriculture budget that afternoon, either echoing the House's decision or opting to renew funding. Working in our favor was the fact that the ensuing battle to reconcile the two budgets could stretch on for weeks, even months. We needed to find support from as many members of Congress as we could.

Dressed in my most convincing suit and a pair of black heels, I grabbed an umbrella from the hall closet before shutting the oak door of my quiet two-story home in the DC suburb of Silver Spring, Maryland, and hurrying to the nearest Metro stop. As the rain blew at me from all directions, I tried not to read the incoming storm as a sign of what the day might bring. One might expect a preacher's daughter would be free of such superstitions. Of course, one might also expect that a preacher's daughter wouldn't be a scientist and a devout believer in evolution. But with annual visits to my grandfather's farm in East Texas, it didn't take me long to realize that God was, at the very least, not the only one contributing to our welfare and survival.

As I settled into a seat on the Metro, I began to review my documents for the morning's briefing. Normally, Kellye would have joined me for the meetings we had scheduled with Senators Dianne Feinstein (D-CA), Carl Levin (D-MI), and Debbie Stabenow (D-MI), and Representatives Mike Thompson (D-CA) and Sam Farr (D-CA), but she was out of town for a conference on wheat genomics. Nervous, but also excited about the opportunity, I paid little attention to the influx of passengers around me.

Thankfully, the L'Enfant Plaza Metro stop is directly connected to the hotel where the scientists had spent the night, so I could avoid the rain. I spotted the four scientists in the hotel lobby right away, having met each of them over the years at meetings of the American Phytopathological Society—a professional organization for plant pathologists. After we exchanged greetings, I offered a short rundown on congressional meeting etiquette:

Do stay on topic.

Don't fall asleep in a meeting.

Don't take a phone call during the meeting.

While these seemed like obvious instructions, I had seen each of these rules violated in previous meetings on the Hill. On one infamous occasion, during a long day of back-to-back meetings with congressional staffers and agency officials, a senior scientist in our group had begun snoring in a slightly-too-warm conference room in the USDA Whitten Building, waking only after multiple throat clearings and nudges from the rest of us. Wanting to make sure they made the biggest impact possible on their visit, the NPDN scientists listened earnestly as I reviewed the protocol, and then, briefcases in tow, we stepped out into the rain and piled into a taxi heading for the National Mall to inform our elected representatives about the importance of plant diagnostic science.

***

Every day, in fields and forests across the nation, plants are under attack by pests and pathogens. Thousands of plant pathogenic microbial species and insect pests are native to the United States, and many more are invasive, entering the country through trade and bearing the potential to quickly decimate large portions of the nation's 914 million acres of farmland as well as natural environments unprepared for their advances. The Mediterranean fruit fly, the Asian citrus psyllid, sweet orange scab, chilli thrips, rice cutworms, plum pox, and Asian soybean rust are only a few of the killers that American farmers have dealt with over the past decade. My client, the NPDN—which coordinates laboratories and experts nationwide and offers training and education in plant disease diagnostics and crisis response—has helped avert the spread of dozens of high-priority plant pests and pathogens, ensuring the safety of our food supply, protecting the livelihoods of the more than two million people who work in the agriculture industry, and saving millions of taxpayer dollars in the process. And yet, despite its successes, the program's $4.4 million annual budget was now at grave risk of being cut, putting America's fields and forests—not to mention the nation's $742.6 billion agricultural industry—in danger.

In the thirteen years I spent training and working as a plant pathologist, I spent many days traversing Texas in a pickup truck to participate in farmer diagnostic training sessions; traipsing through agricultural fields from California to Maryland looking for spots, streaks, specks, wilts, cankers, lesions, and ooze; or running experiments in laboratories to learn how microbes attack plants at the cellular level and how best to stop them. Back when I was grinding up plant leaves to extract DNA, placing bits of plant tissue on Petri plates to observe microorganisms living therein, and studying the genomes of bacteria that cause plant disease, I certainly would not have predicted that one day I would trade in my dusty Dr. Martens and lab coat for a polished wardrobe more suited for meetings with members of Congress. But I had come to learn that science works in mysterious ways. Brilliant minds in laboratories everywhere can work to improve the science of plant diagnostics, but without government backing, the majority of these successes can't be implemented. The first step toward getting the backing that scientific programs need is making people aware that plant diagnostics is much like human diagnostics. When it comes to infectious diseases, whether hosted by humans or plants, it is much easier to prevent a widespread outbreak than it is to stop one that has already begun. So when a post-doctoral fellowship turned into an opportunity to work in the nation's capital, spreading awareness about outbreaks and prevention, I jumped at the chance.

As the taxi driver wove through the steady stream of traffic headed in the direction of the National Mall, I thought about the various pathogen outbreaks I had studied over the years. I wondered how many of these stories were known by members of Congress. The story of Florida's most recent battle with citrus canker, which began several years before the NPDN's founding, was legendary among plant pathologists. But how many policy makers outside of the state of Florida knew about it?

It started on September 28, 1995, when Louis Francillon, a forty-one-year-old agronomist for the Florida State Department of Agriculture, stood outside a house on Southwest 16th Street in Miami-Dade County, performing his routine inspections. Sweating in the afternoon heat, Francillon carefully inspected a triangular flytrap hanging from a branch of a sour orange tree. He was looking for Mediterranean fruit flies—exotic pests that cause extensive damage to a variety of fruit crops, including citrus. No fruit flies were in the trap. As he turned away, however, a nearby grapefruit leaf, backlit by the sun, caught his eye. In the center of the leaf, surrounded by the bright green glow of chlorophyll, was a brown lesion bordered by a yellow halo.

Francillon pulled his magnifying lens out of his pocket, attempting to shield his eyes from the sun's glare as he took a closer look. He knew instantly what he had discovered. The lesion was a telltale symptom of citrus canker, a disease that had originated in Southeast Asia but had spread through agricultural trade to citrus-growing regions around the world. It had been found in Florida in 1912 and 1986, and had been successfully eradicated both times. This time the state would not be so lucky; by the time of Francillon's discovery, the pathogen had already begun to spread. This made containment difficult.

For the next seven years, Florida's Department of Agriculture waged war on the bacterial invader in an effort to protect the state's $9 billion citrus industry, but the disease persisted. Many growers went out of business. Others let go the majority of their employees. With the economy suffering and the threat of statewide crop damage looming, the legislature came up with a solution: cut everything down. Every citrus tree within 1,900 feet of an infection had to go. By 2002, crews were traversing the state, felling trees at a pace that peaked at five thousand per day. In all, 16.5 million trees in groves, nurseries, and backyards were destroyed—so many trees that, if stacked end to end, they would have reached nearly one third of the way to the moon.

But despite these extreme measures, it was too late. The disease had been spreading for nearly a decade when Mother Nature, in the form of a series of hurricanes, carried the disease well beyond infected groves. With all hopes of containment now thwarted, the $1.6 billion eradication program—the largest plant disease eradication program in U.S. history—was ended in 2006.

A more recent disaster came in the form of a small insect. The emerald ash borer—a flat-headed, metallic-green insect, native to Asia and Eastern Russia but found outside of Detroit in 2002, the same year the NPDN was created—is just one of many dangerous "hitchhikers" that have slipped into the United States in wood-packaging materials. Appropriately shaped like a small bullet, the ash borer has killed nearly a hundred million ash trees in eighteen states, from Kansas to Maryland, sweeping through urban areas and suburbs, where ash is commonly planted. As with citrus canker, eradication efforts led to wide-scale tree removal. In some neighborhoods, every tree was taken out in one fell swoop, with only rows of stumps and sawdust left behind. A recent study published by the *American Journal of Preventive Medicine* connected this rapid depletion of the natural environment with increased rates of death from heart attacks and lung disease.

Across the fifteen states in the study area, researchers drew connections between tree loss caused by the emerald ash borer and more than 20,000 deaths.

The NPDN was established to mitigate the impacts of these and other pest and pathogen attacks. Although the program was not created in time to curb the ash borer's initial devastation, it has since developed training materials to prepare scientists, gardeners, and others to identify signs of emerald ash borer infestation. The first detection of the pest in Minnesota, in 2009, was a direct result of NPDN's training program. This early detection led to a state quarantine that has thus far been effective in protecting Minnesota's trees.

Had the NPDN been established prior to citrus canker's spread, Florida's story could also have unfolded differently. With a cadre of NPDN-trained first responders, perhaps the disease would have been spotted even before Francillon's identification, leading to an earlier and more focused response. It's hard to say what might have happened. But the NPDN was put into place in order to prevent new outbreaks from having similar consequences. And yet, the program's successes in doing so had apparently gone unnoticed—a catch-22 that those of us working in agriculture advocacy often face.

The U.S. agriculture industry has largely been successful. It has kept our nation's grocery stores stocked with a ready supply of food, a state of affairs that most of us take for granted. And without looming threats of food shortages or compelling stories, plant and agriculture programs all too frequently take a back seat to the numerous other activities with which they compete for federal funding. After all, the human-interest aspects of many other programs are often more readily apparent. For example, all of us, including members of Congress, have been sick, and most of us have also lost loved ones to illness. Because we can all relate personally to the need for medical advances, convincing voters and Congress to fund medical science is not a particularly hard sell, and, therefore, funding for the National Institutes of Health (NIH) not only remains secure but rises almost annually. Compared to the impacts of cancer and other diseases on our everyday lives, it's easy for plant diseases like citrus canker to seem insignificant. And so, over the past fourteen years, while NIH funding has risen steadily, from $22.9 billion in 2000 to $28.5 billion in 2013, funding for U.S. Department of Agriculture (USDA) science programs, already a small percentage of the amount NIH receives, has declined from $2.4 to $2.2 billion. But despite this much smaller level of funding, agricultural science continues to be responsible for many crucial successes on which our food supply, our jobs, and even our lives depend. I hoped the NPDN scientists and I would be able to remind Congress of this important fact.

***

"Hello, are you the Plant Network?"

Still glancing at his schedule, where the name of our group was printed, a man with a dark goatee and receding hairline entered the reception area where the five of us sat awaiting our 1:30 meeting.

"I'm Troy Phillips, Congressman Sam Farr's Senior Legislative Assistant."

"The National Plant Diagnostic Network, or NPDN for short," I interjected, rising quickly to my feet and extending my hand to begin a round of handshakes and introductions. As we followed Troy into Congressman Farr's office, settling into plush chairs surrounding an ornately carved coffee table, I began by expressing our gratitude for the Congressman's support.

"We would like to thank the Congressman for his commitment to promoting agriculture research," I said.

Although every meeting begins with such pleasantries, I sometimes deliver them through clenched teeth, politely "thanking" members of Congress for their support when in reality they've done quite the opposite. This time, I meant it. Congressman Farr is the Ranking Member of the Appropriations Subcommittee on Agriculture, Rural Development, and Food and Drug Administration, and I was happy that in this meeting, we would likely find a sympathetic audience. It was our third meeting of the day, and we were all beginning to get a little edgy as we waited to hear the Senate's version of the agriculture budget.

"The main reason for our visit today," I continued, "is to provide you with information about the NPDN and answer any questions you may have." Then I jumped to the point: "Funding for the program has been zeroed out in the House budget," I said, motioning the way a baseball umpire calls a runner "safe." I glanced over at NPDN's executive director, Rick Bostock, a UC Davis expert in canker, brown rot, and other fungal fruit diseases. Right on cue, he began to describe the program's significance.

"During an outbreak of plant disease, a timely response is critical," he began, in the practiced tone of a seasoned professor. "Two things need to happen: you must identify the 'positive' plants and clear the 'negatives.'" (Echoing the language we use to talk about human viruses and other infectious agents, "positive" plants refer to those that have been infected by a certain pathogen; "negatives" are the plants that have not.) "If the NPDN is not around to help with that process, retailers may go out of business waiting to be cleared to resume plant sales," Rick explained, referring to the USDA-mandated sales freeze that is levied on plants suspected of harboring particularly worrisome pathogens. "Or, even worse, the outbreak may go unchecked with no NPDN-trained 'first responders' to report it," he added.

I looked across the table at Troy, who reminded me of a younger and better-looking Billy Bob Thornton, searching for signs that Rick's words were making an impact.

Each day, Troy and other legislative aides, senators, and representatives hear hundreds of requests from constituents and the more than 12,000 registered lobbyists in DC. The lobbyists—along with countless special interest groups, like the National Education Association, Citizens United, and AARP, as well as lesser-known groups, like the U.S. Association of Reptile Keepers, the Balloon Council, Cigar Rights of America, and the American Pyrotechnics Association—regularly descend on Capitol Hill like starving persons at a soup kitchen.

Some of these groups let their money do the talking. Take the National Rifle Association, for example, which spends more than $3.4 million per year on lobbying efforts and campaign contributions, or the American Civil Liberties Union, which spends $2 million. But smaller groups, like the American Phytopathological Society (APS), can't rely on endorsements to sway congressional opinion in their favor. Such groups have to rely on their membership—plant pathologists, in the case of APS—to voice their concerns to Capitol Hill.

I continued to watch Troy's reaction as Jim Stack, a regional NPDN director and specialist in the diagnosis and management of field crop diseases in the Biosecurity Research Institute at Kansas State University, took his turn. Jim emphasized the marked improvement in crisis response established by the NPDN. "Prior to the NPDN, information on new pests and diseases was not easily accessible, communications on new outbreaks were poorly coordinated and inadequate, and infrastructure supporting plant diagnostics in the country had degraded," he explained. Before the NPDN was established, scientists in Nebraska, for example, may or may not have been aware of a disease discovery in neighboring Iowa. And training in plant disease diagnostics was limited to a handful of experts, making it far more difficult to identify and combat diseases before they were already established.

Jim leaned forward from his blue leather chair and slid an informational packet across the table to Troy. "The NPDN is an example of a government *success*," he stated. The word lingered in the air like pollen in springtime as we waited for Troy's response.

While the media often capitalizes on disasters like the emerald ash borer, or Florida's battle with citrus canker, agricultural programs also *stop* many disasters from happening. For every pathogen or pest that slips through the system, many more are prevented from spreading—but these kinds of stories rarely make the headlines. Can you imagine what this would look like?

"Plum Pox Found in Routine Inspections. Nothing Happened."

"Beetles Stopped at USDA Checkpoint. Invasion Averted."

I might read these stories. But I'm not sure many other people would or that I would blame them if they didn't. When agriculture is successful, to the eyes of the public, nothing seems to be happening. Crops keep producing, trade ensues, and we continue to have food on our plates. In the absence of a disastrous outbreak, what is there for anyone to notice?

But in this era of endless budget cuts, we couldn't allow NPDN, and its important work, to remain unnoticed. We had to make a case that Troy would remember and pass on to the Congressman. In my time working in advocacy, I have learned that the fastest way to a member of Congress's heart is by demonstrating the impacts of a given program on his or her constituents. Congressman Farr's district included 88,000 farms and ranches nestled in California's central valley and fertile coastal regions and $43.5 billion in agricultural sales in 2011. His support could go a long way toward ensuring the program's future, and we had come prepared with stories about NPDN's impacts on the state.

"You're familiar with sudden oak death?" Jim asked. Troy nodded his head "of course." The disease, characterized by oozing cankers on trunks, foliage dieback, and, eventually, the death of infected trees, had heavily impacted California's stately oaks. In 1995, the microbial menace passed through coastal California forests, attacking the many native oak species, disrupting the natural ecosystem, and leaving swaths of ugly brown tree skeletons in its wake. I knew this was our chance to make the importance of the NPDN hit home. Jim began to tell the story.

In March of 2004, biologists from the California Department of Food and Agriculture, conducting routine inspections at several large wholesale nurseries in southern California, discovered that their camellias and rhododendrons were infected with a fungus-like microbe, *Phytophthora ramorum*—the pathogen that causes sudden oak death. Like its cousin, *Phytophthora infestans,* which wiped out potato crops in the mid-1800s and contributed to more than one million human deaths in Europe, *P. ramorum* is a serious threat—and it doesn't just infect oak trees, but rather is spread through a number of hosts. I could tell by the light in Troy's eyes that Jim's story had captured his interest.

"By the time of the discovery, plants that had likely been exposed to the pathogen had already been shipped throughout the country," Jim explained. This was a serious problem; *P. ramorum* is a "quarantine" pathogen—a blacklisted microbe that triggers U.S. government protocols—and thousands of retailers throughout the country were required to cease business operations. While the jobs of thousands of owners and employees—not to mention the country's oaks and the habitats to which they contribute—all hung in limbo, retailers sent more than 100,000 samples to the USDA Animal and Plant Health Inspection Service (USDA-APHIS) laboratory in Beltsville, Maryland, to be analyzed for signs of infection. This caused an even bigger problem. The scientists in the APHIS diagnostics laboratory can process only twenty to twenty-five samples per day.

Jim paused for impact, vertical lines shooting up his forehead from his furrowed brow. "This was when the NPDN got involved," he explained to Troy. In an effort to avert a nationwide outbreak of sudden oak death, the NPDN participated in an unprecedented investigation, which utilized already existing diagnostic laboratories at universities across the country to speed up a clearance process that would have otherwise resulted in costly downtime for businesses nationwide. It is estimated that by enabling many nurseries to quickly resume operations, the NPDN saved businesses a collective $20 million. Meanwhile, this intervention allowed APHIS laboratories to focus their energies on the positive samples and their primary objective—preventing the spread of sudden oak death.

Jim sat back in his chair as the other scientists described how the NPDN has since been involved in similar investigations of other pathogens, including plum pox virus and Asian soybean rust, with equally impressive results. And yet, despite these impressive successes, NPDN's $4.4 million had been zeroed out in the House version of the 2012 congressional budget. A look of astonishment crept across Troy's face as we continued to offer information on other would-be instigators of disaster and the millions of American jobs that are supported by pest- and disease-free agriculture exports.

"Why would we cut this program?" Troy asked, genuine surprise in his voice as he flipped through the datasheets and pamphlets Jim had given him. He had never heard of the NPDN before, he admitted, expressing his regrets that he didn't have this information when he was fighting for funding in the House.

I glanced anxiously at the clock over the door. It was already a few minutes past 2 p.m. The Senate was probably revealing its version of the 2012 agricultural budget as we sat there. The day's meetings—as well as all of the time and resources invested in NPDN over the past decade—would be a lost cause if the budget was ultimately settled and the program's funding cut.

"In this time of guerilla warfare, we have to hope for our colleagues in the Senate to help us out," Troy lamented. "Who have you talked to in the Senate?" he asked, adjusting his silk tie. I described the group's appointments that day with Senators Feinstein, Levin, and Stabenow, and promised to keep Troy posted as we said our goodbyes.

The group filed out of Congressman Farr's office, stopping in the hallway to assess the meeting. "I think he gets it," Rick said. I nodded as I glanced at my schedule and then back at the clock. We still had two more meetings. We could only hope that the other members of Congress and their aides would be as receptive as Troy. But what if we were too late? What if the Senate had already decided to cut the program? Wistfully, I remembered the quiet of the laboratory, where the only fate that directly hinged upon my actions was that of the specimens in my Petri dishes. But as we shuffled down another long hallway to yet another meeting, I received a text from Kellye: NPDN funding had been restored in the Senate budget. Breathing an audible sigh of relief, I excitedly shared the news with the others.

"Great! We can all go home," one of the scientists joked. We laughed, some of the day's stresses beginning to subside.

I knew, however, that the fight was far from over. The budget would come back for conference, at which point NPDN funding could be nixed again. But I had a feeling that Troy would remember our story. He seemed convinced that the NPDN was a program worth saving. Perhaps I should have been content with this, but I couldn't help but wonder about the fate of other agriculture programs for which Eversole Associates also advocates, ones that lend themselves even less well to a crisis-oriented story. A story of a disease that flattened the nation's 80 million acres of corn crops would necessarily make headlines and turn policymakers' heads. Even Demeter and Chicomecoatl would probably roll over in their mythological graves. But when agricultural science works and agricultural pathogens are kept out of the fields, does anyone really notice? We simply expect this, don't we?

***

Two years after our trip to the Hill, the NPDN's funding remained intact—albeit at an annual level half that of its first eight years of existence. And as I placed my overstuffed tote bag on the conveyor belt at the USDA Whitten Building security checkpoint to escort Jim Stack and two other NPDN colleagues to another round of meetings, I felt lucky to have perhaps played a small part in getting the program off the chopping block. But the bigger question still lingers in my mind: who really thinks about plants? Or, perhaps more importantly, who thinks about the science on which our plant production depends?

I am happy that many U.S. agricultural stories are stories of success and that newsworthy disaster stories, like the emerald ash borer and citrus canker, are few and far between. But this may not always be the case. An increasingly global economy means that pathogens and pests are inadvertently shipped all around the world. It also means that the United States and other countries have become codependent for agricultural commodities; thus, disasters that happen elsewhere, like famines in developing countries or the ongoing coffee rust crisis in Central America, can impact us here at home. Sometimes the impacts on the United States are great—the Irish Potato Famine, for instance, resulted in mass migration that forever changed U.S. demographics. Other times, they seem small—like Central American outbreaks of coffee rust that add a few extra cents to the price of your morning latte. As humans, we depend on plants; by extension, we depend on plant science to ensure that our forests are healthy and our crops are productive. It is incumbent upon those of us working in plant science and advocacy to communicate the importance of our work to policymakers and citizens.

And yet, even were we all to begin our days with expressions of gratitude for plants and plant science, this alone would not be enough. Plant science needs to transcend university walls and research laboratories, and travel beyond the pages of scientific journals and textbooks. A recent study in *Physics World* found that the average journal article is read by only three people. Clearly, the gap between scientific output and public awareness is immense. Closing this gap will require scientists continually to remind policymakers and the public that the science that goes into agricultural production is far more complex than the old adage "eat your vegetables" might imply.

But scientists generally prefer to focus on, well, science. And I don't blame them. The joy of discovery, the satisfaction of proving a hypothesis others never considered, the pleasure of sharing your findings with respected experts in your field—it's easy for this to feel like enough. While working in the laboratory, I frequently danced down the hallway, proclaiming victory after an experiment revealed interesting results. But these same results don't send members of Congress dancing down Capitol Hill's hallways. In fact, too often—as was the case with the NPDN prior to our visits—scientific successes, in particular, go unnoticed. For this reason, it is critical that scientists maintain a dialogue with policymakers so that important programs do not fall through the cracks. Public policy provides the framework for decision-making. But informed policy is what ensures that we continue to make the right decisions.

Clearly, not all plant scientists can make the decision I did, swapping in their field boots and lab coats for a career on the Hill. Nor should they. We need people in laboratories, investigating viruses like wheat streak mosaic, which can be devastating to winter and spring wheat crops, and bacteria like *Pectobacterium carotovorum*, which causes rot in a wide range of plants including potato and tomato. The reality, though, is that more plant scientists should also focus on making others aware of the larger fates that hang in the balance of these findings. While the effects of plant disease may be readily visible to farmers who lose their crops or agriculture workers who lose their jobs, plant pests and pathogens affect all of our lives, sometimes—as in the case of the emerald ash borer— even changing the very air that we breathe. Plants themselves are living, breathing entities on which the rest of our lives depend.

Some people attribute the bounty of the Earth to God's miracles. Fair enough, I suppose. But as the Nobel Prize-winning plant pathologist Norman Borlaug put it, "There are no miracles in agricultural production." The success of U.S. agriculture depends on us all—farmers and workers in the fields, scientists in the laboratories, decision makers on the Hill, and their constituents across the country. Finding ways to make all these groups work together might, indeed, be the real miracle. And it's a necessary one.

6

DROWING IN DATA

Gwen Ottinger with Rachel Zurer

I was at the most undignified moment of moving into my new office—barefoot and on tiptoes on my desk, arranging books on a high shelf—when one of my fellow professors at the University of Washington-Bothell walked in to introduce himself. Pulling my shirt firmly over my waistband, I clambered down to shake his hand and exchange the vital information that begins academic acquaintanceships: Where had I come from? What kind of research did I do?

I felt my shoulders tense, bracing for the question I knew was probably coming next. I explained that I studied communities living next to oil refineries, especially how residents and refinery experts make claims about the effects of chemical emissions on people's health. My colleague replied with what I'd been hoping he wouldn't: "But is it really the emissions from the refineries that are making those people sick?"

An important question, to be sure—essential, even, to policymakers deciding how refineries and petrochemical plants ought to be sited and regulated. So it's hardly a surprise that in the decade since I started my research, I've been asked The Question scores of times, in settings that range from conference presentations to New Orleans dive bars. Yet it's a vexed question, and I have always been frustrated and often struck dumb with my inability to answer it. "There's a lot of controversy over that," I explained to my colleague in my best anthropologist-of-science manner. "The truth is that we don't really know enough to say for sure."

But as I returned to the solitary work of shelving books, I sought refuge in a place that had recently become my favorite environmental fantasy: A brown, windswept hill at the edge of a refinery in the San Francisco Bay area, topped by a small white trailer the size of a backyard tool shed. In my imagination, the trailer glows in the California sun as the state-of-the-art monitoring instruments inside it hum and flash, measuring minute by minute what's in the air. In my imagination, a cadre of scientists peers at computer screens to turn these data into a more satisfying answer to The Question, an answer that matches real-time chemical concentrations with the health concerns of people living nearby.

My fantasy is set in a real place, though I've never seen it. The hill of my imagination overlooks the town of Benicia, a bedroom community of 30,000, where people who drive tight-lipped to San Francisco jobs all week stroll past the antique shops to First Street for scones and lattes on Saturday morning. It's a charming place, yet Benicia's industrial past persists; a slim smokestack pokes up like a flagpole beyond the trailer, its white plume meandering off toward the Carquinez Strait. Benicia is home to one of the 150 or so oil refineries that feed the nation's appetite for energy. Less than a mile from downtown, an Oz of tanks and towers on 800 acres churns away day and night, turning up to 170,000 barrels of oil per day into gasoline, asphalt, jet fuel, and other petroleum products. The Valero facility is the town's biggest employer and the major denizen of Benicia's industrial park. The trailer sits on its southern edge.

Most of the communities I have studied are clustered in the South and are smaller, poorer, and more economically dependent on their refineries than is Benicia. For them, the trailer and the data it offers are even more urgent than they are for Benicia residents. These "fenceline communities" are places where people cough. Where they carry asthma inhalers. Where every resident has a handful of neighbors who have died of cancer. Where refinery and government officials insist that chemicals in the air don't harm them, and residents are sure that they know better. These communities are places where conflict lingers in the air along with the smell of sulfur.

Data that can show how chemical exposures are related to health symptoms could help these communities. It could suggest the kinds of protection they need, could show the real extent of emissions reductions necessary on the part of the refineries, could point the way to improved environmental policies. In my mind, Benicia's trailer gleams with the possibility of new knowledge that helps everyone.

But a few weeks after my colleague's visit, my hopes for the trailer dimmed. As I was putting the finishing touches on a syllabus in my office, by now already messy, the phone rang. It was Don Gamiles, an engineer whose company installed Benicia's trailer. He had been excited about the project in Benicia from the time he first mentioned it to me earlier in the summer.

Gamiles has been involved in air monitoring since the aftermath of the Persian Gulf War, when he ran equipment to detect potential poison gas releases during United Nations inspections of Iraqi facilities. He's invented two instruments that can measure concentrations of toxic gases in real time, both of which are part of the suite of monitors that he pulled together for the trailer in Benicia. But these days, Gamiles's business really centers on mediating conflicts between facilities that release those gases and neighboring communities concerned about them. Affable and unassuming in his characteristic polo shirt and khakis, Gamiles works with both sides to design and install suites of monitors, like the one in Benicia, that incorporate his instruments and produce solid data about what's in the air so that neither side can exaggerate. "Everyone's a little bit right," he says. "The refinery guys tend to over-trivialize what's coming out. But communities want to make them the villain."

Though he's been involved in other projects (one major refiner is even talking about making Gamiles's monitors a standard part of their environmental best practices), the Benicia project is what Gamiles raves about: "The sampling station's the best in the world," he said, reminding me that it can monitor hydrogen sulfide, black carbon, and particulates in addition to hazardous air pollutants such as benzene, xylene, and toluene, all for a very reasonable price tag. And the best part: "Everybody's happy!" He chuckled and I imagined his self-effacing grin. "This is a model of how to do things right."

"There's just this one sticking point," he added. He'd called to ask for my help. The refinery and the community group that pushed for the monitors were having trouble figuring out how to present the data. If the monitors detected chemicals, how could they best explain what that meant to someone looking at that data on a public Web site?

The refinery, it seemed, wanted to avoid alarmism and irate hordes at their gates; on the other hand, it was in no one's interest if they swept real risks under the rug. "Everybody has a valid point," Gamiles said. "What would be helpful to have is a listing of standards for all of this stuff"—all of the chemicals that the monitoring station could be detecting, starting with benzene, toluene, xylene, and sulfur dioxide. Could I work with a student to put together a list?

My heart sank. Here was The Question again, in a more nuanced form. Gamiles was asking, "At what exposure levels do emissions from refineries make people sick?" Worse, this wasn't the first time I'd been asked to take stock of the available information, and what I'd found the last time had driven me to my fantasies of fancy new monitors in the first place.

Buckets of Data

In the summer of 2001, I was halfway through my 20s and a Ph.D. program when I walked into the Oakland, California, offices of a nonprofit organization called Communities for a Better Environment (CBE). After years with my nose in a book, I was dying to do something "real" and antsy about finding a focus for my thesis project. I hoped that interning for CBE, whose lawyers, scientists, and organizers worked with Northern California communities to advocate for environmental justice, might address both problems at once.

No one was at the reception desk, so I hung by the door, fingering pamphlets and newsletters announcing the organization's latest successes, including its work helping refinery-adjacent communities establish "bucket brigades" to monitor air quality with do-it-yourself air samplers made from hardware store supplies. Eventually someone bustled past and directed me to the Science Department at the end of one of the office's warren-like hallways.

My first assignment seemed simple enough: Communities were getting data with their bucket samples, but they were having a hard time saying what the numbers meant. My job was to compile a list of the state and federal air standards for different chemicals that might show up in a bucket sample. The list would be like a yardstick that citizens could use to put air quality readings in perspective, showing how the numbers measured up to the thick black line that separated "safe" from "dangerous."

As a starting place, my supervisor handed me a second-generation photocopy of a fax containing a table of numbers. The fax was from Wilma Subra, a MacArthur "genius grant"-winning chemist and legend among refinery-adjacent communities in Louisiana. Subra's document listed "levels of concern"; specifically, the regulatory standards set by Louisiana and nonenforceable "screening level" recommendations from the neighboring state of Texas. I was to expand the table, adding comparable standards from other agencies, to give bucket users a straightforward way to know when the concentrations they measured were cause for alarm.

Squinting at a computer screen from the corner of a borrowed desk, navigating through one agency Web page after another in search of air quality standards, I had no problem adding columns to Subra's chart. Agencies such as the Louisiana Department of Environmental Quality (LDEQ), its counterparts in Texas and North Carolina, and the American Toxic Substances and Disease Registry set standards or made recommendations for acceptable ambient air levels of individual chemicals. But each included only a subset of the chemicals I was looking for. The federal Clean Air Act, for example, set limits on total volatile organic compounds, a category that includes these chemicals, but not on the individual air toxins under that umbrella, such as benzene, toluene, and xylene: monoaromatic hydrocarbons known or suspected to cause cancer.

As the table grew, I was surprised to find that there was no consensus on what constituted a safe or permissible level for any of the chemicals. Even after I'd converted the disparate standards into a common unit of measurement, reading across any one row (for benzene, say, or hydrogen sulfide), there were numbers in the single digits, in the double digits, decimal numbers. The lack of consensus was apparent even in the table's header row: One agency set limits on 8-hour average levels, the next on annual averages, the next on 24-hour averages. There didn't even seem to be agreement on what period was most appropriate for any given chemical. I didn't have a single yardstick; I had several of them, each for a different kind of measurement, each with multiple black lines. How would this help anyone figure out what chemical concentrations they should worry about?

At my boss's urging, I made some phone calls to find out how the agencies could arrive at such different standards. A scientist at the LDEQ explained that his agency used occupational health studies—studies of how workers were affected by the chemicals—and multiplied the results by a scaling factor. I remembered the number from my graduate class in risk analysis: it adjusted risk levels based on 8-hour-a-day, 5-day-a-week worker exposures to numbers appropriate for populations such as people living near refineries that could be exposed to the same chemicals for as much as 24 hours a day, seven days a week.

A Texas regulator, in contrast, told me that her agency based its recommendations mostly on laboratory studies. I knew about this process from my class, too. Groups of mice or rats or other small animals would be exposed to varying levels of a chemical to determine the highest dose at which the animals didn't appear to suffer any adverse health effects. The agency scientist would have looked at a number of different studies, some of them with incompatible results, made a judgment about which numbers to use, then applied a safety factor in case human populations were more sensitive to the chemical than other mammals. But what neither she nor her counterpart in Louisiana had to work with were studies of what these chemicals did to people who breathed lots of them at a time, in low doses, every day.

In the end, digging into the standards and learning how incomplete and uncertain they were convinced me that we don't really have a good answer about exactly what the chemical levels mean for health. Anyone who professes to know with certainty is operating as much on belief as on data. So by the time Don Gamiles asked me, nine years later, if I could assemble the standards for the chemical that his shiny new monitoring station was detecting, I wanted to tell him that all he was going to get was a whole bunch of yardsticks. What he needed was an additional stream of data, health data that could put chemical concentrations in the context of real people's experiences and, over time, help put those standards on a firmer footing.

But Gamiles is an engineer, not an epidemiologist. I knew that his contract would not have funding for what I was proposing. And explicitly mentioning the health concerns wasn't likely to help Gamiles maintain the collegiality between the Valero refinery and its neighbors in Benicia.

I took a deep breath and agreed to look for a student who would investigate the standards. Maybe, I told myself, if we could show Gamiles and the engineers at Valero the uncertainties in the standards, we could start a richer conversation about what the data coming from the new monitoring station meant, and how to figure it out.

Having that conversation, or at least trying to, seemed especially important since more and more refineries, especially in environmentally conscious parts of the country such as the San Francisco Bay area, have been seeking

Gamiles's services, installing their own monitors before an increasingly vigilant Environmental Protection Agency (EPA) can require them to. And yet part of me knew that imagining I could get refiners and communities to talk about the issue was overly optimistic, if not downright naïve. I already knew that petrochemical companies weren't troubled by the limitations of the standards. In fact, years earlier in Louisiana, I'd seen how they use those very uncertainties and omissions to their advantage.

The Lowdown in Louisiana

Many of the air monitors in the trailer in Benicia hadn't yet been developed when Margie Richard decided to take on the Shell Chemical plant across the street from her home in Norco, Louisiana, in the late 1980s. But what was in the air, and what it could do to a person's health, were very much on her mind.

Richard's front windows looked out on an industrial panorama: tall metal cylinders and giant gleaming spheres connected by mazes of pipes, all part of the processes that turn crude oil into gasoline, ethylene, propylene, and industrial solvents. Half a mile away, at the other edge of the 3,700-person town, an oil refinery loomed. On good days, a faint smell of motor oil mixed with rotten eggs hung in the air; on bad days, chemical odors took Richard's breath away.

Throughout Richard's eight-square-block neighborhood of Diamond, the historic home of Norco's African-American population, people were getting sick. Richard's young grandson had asthma attacks that landed him in the emergency room on more than one occasion. Two streets over, Iris Carter's sister died in her forties of a disease that doctors told the family they only ever saw in people living near industrial facilities.

Barely five feet tall and bursting with energy even in her early sixties, Richard led her neighborhood in confronting Shell about its plant's ill effects. Every Tuesday afternoon, she and a few other women with picket signs walked up and down the far side of her street, in front of the chain link fence that separated Shell from the community, demanding that representatives from the company meet with residents to discuss a neighborhood relocation. Concerned about their health and safety, she and other residents wanted out.

By 1998, Richard and her neighbors finally started to get some quantitative data to support their claims that Shell's emissions were making them sick. Denny Larson, then an organizer with CBE in Oakland, arrived with buckets. With the low-tech air sampler—little more than a five-gallon plastic paint bucket with a sealed lid and a special bag inside—Richard documented an incident at Shell Chemical that emitted potentially dangerous concentrations of an industrial solvent called methyl ethyl ketone (MEK). She also gathered evidence that residents of her community were exposed to toxic chemicals when odors were inexplicably bad, and even personally presented a high-ranking Shell official with a bag of air from her community at a shareholder's meeting in the Netherlands.

In 2002, Richard and her group triumphed. Shell agreed to buy out any Diamond residents who wanted to leave. But Richard had succeeded in more than winning relocation. She had also put air monitoring on Shell's agenda, where it had not previously been. That fall, even as families in Diamond were loading moving vans and watching bare ground emerge where their neighborhood had been, Shell Chemical and its Norco counterpart, Motiva Refining, launched their Air Monitoring...Norco program.

Good Neighbors

One muggy September afternoon, I picked up a visitor's badge at the guardhouse at Shell Chemical's East Site and made my way to the company's main office building. The rambling, two-story beige-and-brown box could have been in any office park in suburban America, except that in place of manicured gardens and artificial lakes, it was surrounded by distillation towers and cracking units.

David Brignac, manager of Shell's Good Neighbor Initiative, which was overseeing the Air Monitoring...Norco program, greeted me with a boyish grin and a slight Louisiana drawl and led me upstairs to his roomy office. We sat at a small round table with Randy Armstrong, the good-natured but no-nonsense Midwesterner in charge of health, safety, and environment for Shell Norco.

Brignac walked me through a printed-out PowerPoint presentation: Surveys showed that Norco residents thought that there were dangerous chemicals in the air and that they had an impact on people's health. Air Monitoring...Norco sought hard data about what really was in the air.

Scribbling frantically on a legal pad, I noted what he left out as well as what he said. There was no mention of the bucket samples; no suggestion that Shell's decision to relocate Diamond residents may have fueled the perception that the air was somehow tainted; no hint at the regulatory enforcement action, taken in the wake of the MEK release, that required a "beneficial environmental project" of Shell; in short, there was no acknowledgement that the monitoring never would have happened if not for the Diamond community's activism.

Using their pencils to move me through their talking points, the two engineers described how the data produced by the program would be "objective, meaningful, and believable." Brignac described a planning process that had included not only Shell and Motiva engineers, but also state regulators, university scientists, and community members. Armstrong outlined a sampling procedure that replicated the one used by the LDEQ in their ambient air monitoring program: Each sample would be taken over a 24-hour period, on rotating days of the week (Monday this week, Sunday next), and their results averaged together, all to ensure that the data gave a "representative" picture of Norco's air quality and not anomalous fluctuations.

Like all good scientists, Brignac and Armstrong acknowledged that they didn't know what their study would find. They monitored emissions leaving the plant, Armstrong explained, and used computer models to predict how they would disperse into surrounding areas. Those models gave them every reason to believe that the air quality was fine. And the company had done studies of its workers' health, which also gave them confidence that their emissions weren't making anyone sick. But we all knew that models aren't measurements, and the health of adult plant workers may or may not say anything about the health of residential populations that include the very young and very old. So with a slightly nervous laugh (or was that my imagination?), Armstrong assured me that Shell would be releasing the results even if they showed that air quality was worse than they had thought.

Nearly six months later, I followed Margie Richard, now a resident of the nearby town of Destrehan, into Norco's echoey, warehouse-like American Legion Hall. Brignac and Armstrong milled with their colleagues near the table of crackers, cheese, and that unfathomable Louisiana delicacy, the shrimp mold. They greeted us warmly as the facilitator began to usher people to their seats for the presentation of Air Monitoring...Norco's first set of results.

A nervous young African-American man from Brignac's Good Neighbor Initiative team began by explaining the rationale and process of the program, using more or less the same slides that I had seen in September. Then a white 30-something from the independent firm that had carried out the monitoring, less polished than his Shell counterparts and looking uncomfortable in his tie, gave us the results. The headline: "Norco's air meets state standards." They had compared the concentrations measured in Norco, he explained, to limits on chemical concentrations set by the LDEQ, and the measured levels were below the regulatory limits.

Neither the contractor nor the assembled Shell representatives said so explicitly, but the conclusion they wished us to draw was clear: Air quality in Norco met the state's standards, so it was perfectly healthy to breathe. I wanted to object. How could they say that when there were no standards for some of the chemicals that they measured? When Louisiana's standards represented just one version of where scientists drew the line between "healthy" and "dangerous"? I sat on my hands and held my tongue; rabble-rousing at public meetings is not an anthropologist's mandate, especially when she hopes to continue interviewing all sides.

But I wasn't the only one inclined to question the implication that "meets standards" was the same as "safe." In the question-and-answer period, a middle-aged African-American woman, her graying cornrow braids piled neatly in a bun, stood up and asked just how good those standards were. How could we know that they were strict enough? One of the university scientists involved in the project, a public health researcher from Tulane, reassured her that the standards were based on the best available scientific studies and updated as new information became available. Shell's engineers nodded their approval. For them, it seemed, Air Monitoring...Norco had settled the matter: There was no reason to think that emissions from Shell were making anyone sick.

Elsewhere in the audience, Margie Richard pursed her lips. I couldn't tell what she was thinking, but the fact that she was there at all, even after having moved away from Norco, suggested that the Air Monitoring...Norco program had been an important aspect of her group's victory. For years, her group had been calling for hard data about the chemicals they were exposed to, and they had gotten it. But in the drafty warehouse, the victory seemed hollow. Shell had interpreted their data in the context of questionable standards in order to prove what they had believed all along. I wondered if Richard was disappointed. I was.

The story didn't have to end there, of course. Residents of Diamond and other fenceline communities had challenged the industry's science before with their bucket samples. They could likewise have launched an attack on the idea that "meeting standards" was the same as "safe" and insisted on health monitoring to go along with the air monitoring. But their relocation victory meant that Diamond's activists were already scattered to new neighborhoods. Battles over the adequacy of standards were not likely to be fought in Norco.

Yet the question remains for other communities: As more and more facilities set up air monitoring programs to satisfy the demands of concerned neighbors, will community activists continue to push to see that monitoring data are used to get better answers about how chemicals affect their health? Or will they accept comparisons to existing standards that rubber-stamp the status quo? Whether the trailer in Benicia turns out to be the breakthrough I've been imagining it to be rests on what residents do with its data.

California Dreaming

When I talked to Don Gamiles in the fall, I had my own favor to ask of him: Would he talk to my colleague, Oakland-based writer Rachel Zurer, and introduce her to the people he had been working with in Benicia? We were working together on a story about monitoring and wanted to know more about the exemplary collaboration that he was involved in. Valero, it turns out, wasn't ready to talk about the project; perhaps they didn't want anyone wondering why the public didn't have access to the data yet. But Marilyn Bardet, the founder of the citizen's group in Benicia that helped pressure the company to install the air monitoring trailer, was more than happy to meet with Zurer.

On a blustery morning in October 2010, Bardet welcomed Zurer into her manicured bungalow on Benicia's east side, then retreated to her office to finish an e-mail. Zurer was left to nose around in the dining room, where Bardet's dual identities were on display.

Bardet, 62, is a professional artist who seems to spend as much time as a community activist as she does painting and writing poems. The walls, shelves, end tables, and cupboards of the dining room were decorated with paintings, sculptures, and shells. But the wood of the dining table hid beneath stacks of papers and files relating to Bardet's newest project: a bid to help her town qualify for federal funding to clean up an old munitions site in town, money she said that city employees hadn't known to request.

Bardet returned in a few minutes, talking quickly. That afternoon she had a meeting scheduled with some Valero officials to keep working out the details of the air monitor's Web site—trying to work through the problem that Gamiles had brought up on the phone, of how to present the data publicly—and she'd been sending them a last-minute memo reiterating her goals for the project. As she gathered her car keys and led Zurer out the door for a tour, she caught her guest up on the details.

Some Benicia residents don't think about the refinery, Bardet explained as she drove under the freeway, past an elementary school, and turned left and uphill just before reaching the Valero property's southern border. It doesn't fit the image of their quaint, comfortable town, and as luck would have it, the prevailing winds tend to sweep refinery odors away from the people, out to sea. The refinery has a good safety record and no history of major conflicts with its neighbors. From many places in town, it's invisible.

Yet Bardet and her fellow members of the Good Neighbors Steering Committee (GNSC) keep a sharp eye on Valero. Keenly conscious of the toxic problems other fenceline communities such as Norco have faced, they are wary of the industrial giant in their midst. The air monitoring station is a product of their vigilance. In 2008, the company made changes to some construction plans without going through the full environmental review that those changes required. Dexterous in navigating the intricacies of bureaucratic requirements, Bardet and the GNSC used Valero's mistake to require the refinery to pay for environmental benefits in Benicia. A single letter Bardet wrote detailing Valero's missteps, plus many hours of work by the GNSC, netted the community $14 million. The monitoring trailer was part of the package.

Bardet parked the car at the end of a residential cul-de-sac and escorted Zurer to a spot under an ash tree in the vacant lot between number 248 (white picket fence, a baby-blue Volkswagen Bug in the driveway) and number 217 (single-story ranch with gray siding, two boats, and a satellite dish). She pointed toward the minor white bump on the horizon, curtained by tall stalks of thistles atop a small brown hill a hundred yards across an empty field. It was the monitoring station that I'd been conjuring in my imagination since Gamiles first mentioned it.

"You wouldn't know that this is a big deal," Bardet said. And it was true. In person, the trailer looked like nothing special. But back in the car again, through lunch at a restaurant in town, all the way until Bardet zoomed off to her meeting with Valero, Bardet shared with Zurer her vision of what the monitors might mean for her community, and for her future as an activist.

"It's not just the refinery," she explained. She pointed out that, for example, while Benicia's elementary school is less than a mile from Valero, it's also near a corporation yard, a gas station, a highway cloverleaf, and the major road through town. The air monitors and weather station could expose exactly which pollutants are infiltrating the school, from where, and under what conditions.

"With that information, you can give a school district an idea of how to improve their site, so you can mitigate it," she said. Teachers could avoid opening windows during rush hour. Or community activists like Bardet would have the data they'd need to evaluate the effect of a new development that would add more traffic to the road. "Policy needs to be evidence-based," Bardet explained to Zurer. "That's what we're after."

Scientific Realities

Zurer called with her report on her meeting with Bardet as I was answering a flurry of e-mails from students worried about their final papers. Hearing Bardet's vision for the monitoring station, my hopes sank further. It wasn't that they weren't going to use the data; indeed, it seemed that the information that the monitoring station produces will be something that Bardet can leverage in her myriad projects to improve her community. But in her pursuit of evidence-based policy, Bardet takes for granted the same thing that the engineers at Shell did and that Gamiles does. She assumes that she has a yardstick that shows where "safe" levels of toxins and particulates in the air become dangerous ones, and that there are reliable benchmarks that would tell teachers when they should close their windows and city officials when more traffic would be too much.

Maybe my pessimism is ill-founded. Maybe the ongoing struggle between Valero and residents over how to present the data will ultimately open the Pandora's box of questions surrounding air quality standards—how they're set, how good they are, how they could be improved—and convince Bardet that she needs a better yardstick. Maybe an enterprising epidemiologist will be seduced by the vast quantities of exposure data that this monitoring station, and others around the Bay area, are producing and persuade Bardet and her group to institute complementary health monitoring in order to create a better yardstick. Maybe the Centers for Disease Control's National Conversation on Public Health and Chemical Exposures, which acknowledges the importance of environmental health monitoring, will help convince government agencies to sponsor such a study.

Maybe, in the end, it was just the stack of grading on my desk that had sucked my hope away. But despite the piles of new information that Benicia's monitoring station will produce—is, indeed, already producing—I couldn't convince myself that any new knowledge would be made, at least not in the absence of more fundamental changes. I wandered off to the faculty holiday party conjuring a new daydream: The National Institute of Environmental Health Sciences would call for proposals for studies correlating air monitoring with environmental health monitoring; the EPA, making ambient air toxics standards a new priority, would demand that data from fenceline communities be a cornerstone of the process; and Marilyn Bardet would seize on the new opportunities and make her community part of creating a better answer to The Question.

ABOUT THE AUTHORS

Elizabeth Popp Berman

Elizabeth Popp Berman is an associate professor of sociology at the University at Albany, SUNY, and the author of *Creating the Market University: How Academic Science Became an Economic Engine* (Princeton University Press, 2012).

Adam Briggle

Adam Briggle, an associate professor in the Department of Philosophy and Religion Studies at the University of North Texas, is the author of *A Field Philosopher's Guide to Fracking* (Liveright Publishing, 2015) among other books.

Ross Carper

Ross Carper is a writer based in Washington State whose publications include fiction, poetry, and narrative nonfiction.

Roberta Chevrette

Roberta Chevrette is a Ph.D. candidate in the Hugh Downs School of Human Communication at Arizona State University. Her research focuses on rhetoric, cultural studies, gender and sexuality, and public memory.

David Guston

David Guston is a professor in the School of Politics and Global Studies and co-director of the Consortium for Science, Policy & Outcomes at Arizona State University.

Lee Gutkind

Lee Gutkind is the founder and editor of the literary magazine *Creative Nonfiction,* the Distinguished Writer in Residence at the Consortium for Science, Policy & Outcomes at Arizona State University, and the author or editor of more than thirty books.

Gwen Ottinger
Gwen Ottinger is an assistant professor in the Center for Science, Technology, and Society and the Department of Politics at Drexel University. She is author of *Refining Expertise: How Responsible Engineers Subvert Environmental Justice Challenges* (NYU Press, 2013).

Angela Records
Angela Records is a consultant with Eversole Associates, a global science and technology firm based in the United States.

Sonja Schmid
Sonja Schmid is an assistant professor in Science and Technology studies at Virginia Tech.

Meera Lee Sethi
Meera Lee Sethi is a Seattle-based writer who is currently pursuing graduate studies in ecology.

Sara Whelchel
Sarah Whelchel is a medical student at the GRU/UGA Medical Partnership in Athens, Georgia.

Michael L. Zirulnik
Michael Zirulnik is a research associate at the Consortium for Science, Policy & Outcomes and a Ph.D. candidate in the Hugh Downs School of Human Communication at Arizona State University. He is Chair of the Peace and Conflict Communication Division of the National Communication Association based in Washington, DC.

Rachel Zurer
Rachel Zurer is senior content editor at *Backpacker* magazine in Boulder, Colorado.

ACKNOWLEDGEMENTS

The editors of this book would like to thank the National Science Foundation* and our Program Officer, Alphonse DeSena, for supporting *Think Write Publish*—joining us on a successful six-year experiment that has shown merit and public impact.

Thank you to the many individuals who graciously donated their time to work with us over the years, including: Laura Helmuth, science and health editor for *Slate;* Mark Rotella, editor and reviewer of nonfiction for *Publisher's Weekly;* Victoria Pope, deputy editor of *National Geographic;* Christopher Cox, senior editor of *Harper's Magazine;* Scott Stossel, editor of *The Atlantic;* Michael Rosenwald, reporter for the *Washington Post;* Ellen Ficklen, former editor of Narrative Matters for *Health Affairs;* Emily Loose, an independent literary agent; Stephen Morrow, executive editor of Dutton Press/Penguin Group; Leslie Meredith, vice president and senior editor of Free Press/Simon & Schuster; Hattie Fletcher, managing editor of *Creative Nonfiction* magazine; Dan Sarewitz, co-editor of *Issues in Science and Technology;* Stewart Moss and Sunil Freeman, the director and assistant director of The Writer's Center; Dr. Jim Buizer at the University of Arizona;

---

* This project is funded in part by the U.S. National Science Foundation (#1149107). Any findings, observations, or opinions expressed are those of the principal investigators and participants and do not necessarily reflect the views of the National Science Foundation.

professor Gary Dirks at Arizona State University; Dr. Merlyna Lim at Carleton University.

Thank you to Kevin Finneran—editor-in-chief of *Issues in Science and Technology*—for taking a chance on disrupting the status quo to publish—for the first time—creative nonfiction in *Issues,* an esteemed publication of the National Academies.

We'd also like to thank Lori Hidinger, managing director, and Bonnie Lawless, program coordinator at the Consortium for Science, Policy & Outcomes for diligently supporting us behind the scenes. And thank you to Arizona State University, whose structure and support allows us to dream, to create, and to impact society in new and powerful ways.

Finally, thank you to the 48 *Think Write Publish* fellows and six mentors for their years of dedication to this project. Their ideas, thoughtfulness, and rich desire to share complex, specialized knowledge with the rest of us in a way that would simultaneously entertain while inform was the focus of this project.